Contemporary Physical Geography

Contemporary Physical Geography

Abhilash Ghosh

RANDOM PUBLICATIONS

NEW DELHI - 110 002 (INDIA)

Contemporary Physical Geography

ISBN 978-93-51112-66-2

Published in 2017 in India by

RANDOM PUBLICATIONS

4376-A/4B, Gali Murari Lal, Ansari Road
New Delhi-110 002
Phone: +9111-43580356, 23289044
E-mail: randomexports@gmail.com; sales@randompublications.com; info@randompublications.com

Reprinted 2024

Type Setting by: Friends Media, Delhi-110089
Digitally Printed at : Replika Press Pvt. Ltd.

Preface

Historical traditions in geographical research are the spatial analysis of the natural and the human phenomena (geography as the study of distribution), the area studies (places and regions), the study of the man-land relationship, and the research in the earth sciences. Nonetheless, modern geography is an all-encompassing discipline that foremost seeks to understand the Earth and all of its human and natural complexities - not merely where objects are, but how they have changed and come to be. Geography has been called "the world discipline" and "the bridge between the human and the physical science". Geography is divided into two main branches: human geography and physical geography.

From the birth of geography as a science during the Greek classical period and until the late nineteenth century with the birth of anthropogeography or Human Geography, Geography was almost exclusively a natural science: the study of location and descriptive gazetteer of all places of the known world. Several works among the best known during this long period could be cited as an example, from Strabo (Geography), Eratosthenes (Geography) or Dionisio Periegetes (Periegesis Oiceumene) in the Ancient Age to the Alexander von Humboldt (Cosmos) in the century XIX, in which geography is regarded as a physical and natural science, of course, through the work Summa de Geografía of Martín Fernández de Enciso from the early sixteenth century, which is indicated for the first time the New World.During the eighteenth and nineteenth centuries, a controversy exported from Geology, between supporters of James Hutton (uniformitarianism Thesis) and Georges Cuvier (catastrophism) strongly influenced the field of geography, because geography at this time was a natural science since Human Geography or Antropogeography had just developed as a discipline in the late nineteenth century.Two historical events during the nineteenth century had a great effect in the further development of physical geography. The first was the

European colonial expansion in Asia, Africa, Australia and even America in search of raw materials required by industries during the Industrial Revolution. This fostered the creation of geography departments in the universities of the colonial powers and the birth and development of national geographical societies, thus giving rise to the process identified by Horacio Capel as the institutionalization of geography. One of the most prolific empires in this regard was the Russian. A mid-eighteenth century many geographers are sent by the Russian altamirazgo different opportunities to perform geographical surveys in the area of Arctic Siberia. Among these is who is considered the patriarch of Russian geography: Mikhail Lomonosov who in the mid-1750s began working in the Department of Geography, Academy of Sciences to conduct research in Siberia, their contributions are notable in this regard, shows the soil organic origin, develops a comprehensive law on the movement of the ice that still governs the basics, thereby founding a new branch of Geography: Glaciology.

The primary aim of this book is to present a succinct account of the essential features of this subject. It is being prepared by keeping in view the requirements of the students and academic professionals.

I thank all members of my team who have helped in the preparation of the book. My special thanks go to "Random Publications" who have published the book.

—Abhilash Ghosh

Contents

1

Introduction

The discipline of geography has a history that stretches over many centuries. Over this time period, the study of geography has evolved and developed into an important form of human scholarship. Examining the historical evolution of geography as a discipline provides some important insights concerning its character and methodology. These insights are also helpful in gaining a better understanding of the nature of physical geography.

History of Geography and Physical Geography

Some of the first truly geographical studies occurred more than four thousand years ago. The main purpose of these early investigations was to map features and places observed as explorers travelled to new lands. At this time, Chinese, Egyptian, and Phoenician civilizations were beginning to explore the places and spaces within and outside their homelands. The earliest evidence of such explorations comes from the archaeological discovery of a Babylonian clay tablet map that dates back to 2300 BC. The early Greeks were the first civilization to practice a form of geography that was more than mere map making or cartography. Greek philosophers and scientist were also interested in learning about spatial nature of human and physical features found on the Earth. One of the first Greek geographers was Herodotus (circa 484 - 425 BC). Herodotus wrote a number of volumes that described the human and physical geography of the various regions of the Persian Empire. The ancient Greeks were also interested in the form, size, and geometry of the Earth. Aristotle (circa 384 - 322 BC) hypothesized and scientifically demonstrated that the Earth had a spherical shape. Evidence for this idea came from observations of lunar eclipses. Lunar eclipses occur when the Earth casts its circular shadow on to the moon's surface. The first individual to accurately calculate the circumference of the Earth was the Greek geographer Eratosthenes (circa 276 - 194 BC). Eratosthenes calculated the equatorial circumference to be 40,233 kilometres using simple geometric relationships. This primitive calculation was unusually accurate. Measurements of the Earth using modern satellite technology have computed the circumference to be 40,072 kilometres.

Most of the Greek accomplishments in geography were passed on to the Romans. Roman military commanders and administrators used this information to guide the expansion of their Empire. The Romans also made several important additions to geographical knowledge. Strabo (circa 64 BC - 20 AD) wrote a 17 volume series called "Geographia". Strabo claimed to have travelled widely and recorded what he had seen and experienced from a geographical perspective. In his series of books, Strabo describes the cultural geographies of the various societies of people found from Britain to as far east as India, and south to Ethiopia and as far north as Iceland. Strabo also suggested a definition of geography that is quite complementary to the way many human geographers define their discipline today. This definition suggests that the aim of geography was to *"describe the known parts of the inhabited world ... to write the assessment of the countries of the world [and] to treat the differences between countries"*.

During the second century AD, Ptolemy (circa 100 - 178 AD) made a number of important contributions to geography. Ptolemy's publication *Geographike hyphegesis* or *"Guide to Geography"* compiled and summarize much of the Greek and Roman geographic information accumulated at that time. Some of his other important contributions include the creation of three different methods for projecting the Earth's surface on a map, the calculation of coordinate locations for some eight thousand places on the Earth, and development of the concepts of geographical latitude and longitude. Little academic progress in geography occurred after the Roman period. For the most part, the Middle Ages (5th to 13th centuries AD) were a time of intellectual stagnation. In Europe, the Vikings of Scandinavia were the only group of people carrying out active exploration of new lands. In the Middle East, Arab academics began translating the works of Greek and Roman geographers starting in the 8th century and began exploring southwestern Asia and Africa. Some of the important intellectuals in Arab geography were Al-Idrisi, Ibn Battutah, and Ibn Khaldun. Al-Idrisi is best known for his skill at making maps and for his work of descriptive geography *Kitab nuzhat al-mushtaq fi ikhtiraq al-afaq* or *"The Pleasure Excursion of One Who Is Eager to Traverse the Regions of the World"*. Ibn Battutah and Ibn Khaldun are well known for writing about their extensive travels of North Africa and the Middle East.

During the Renaissance (1400 to 1600 AD) numerous journeys of geographical exploration were commissioned by a variety of nation states in Europe. Most of these voyages were financed because of the potential commercial returns from resource exploitation. The voyages also provided an opportunity for scientific investigation and discovery. These voyages also added many significant contributions to geographic knowledge. Important explorers of this period include Christopher Columbus, Vasco da Gama, Ferdinand Magellan, Jacques Cartier, Sir Martin Frobisher, Sir Francis Drake, John and Sebastian Cabot, and John Davis. Also during the Renaissance, Martin Behaim created a spherical globe depicting the Earth in its true three-dimensional form in 1492. Behaim's invention was a significant advance over two-dimensional maps because it created a more realistic depiction of the Earth's shape and surface configuration.

In the 17th century, Bernhardus Varenius (1622-1650) published an important geographic reference titled Geographia generalis (General Geography: 1650). In this volume, Varenius used direct observations and primary measurements to present some new ideas concerning geographic knowledge. This work continued to be a standard geographic reference for about a 100 years. Varenius also suggested that the discipline of geography could be subdivided into three distinct branches. The first branch examines the form and dimensions of the Earth. The second sub-discipline deals with tides, climatic variations over time and space, and other variables that are influenced by the cyclical movements of the Sun and moon. Together these two branches form the early beginning of what we collectively now call physical geography. The last branch of geography examined distinct regions on the Earth using comparative cultural studies. Today, this area of knowledge is called cultural geography. During the 18th century, the German philosopher Immanuel Kant (1724-1804) proposed that human knowledge could be organized in three different ways. One way of organizing knowledge was to classify its facts according to the type of objects studied. Accordingly, zoology studies animals, botany examines plants, and geology involves the investigation of rocks. The second way one can study things is according to a temporal dimension. This field of knowledge is of course called history. The last method of organizing knowledge involves understanding facts relative to spatial relationships. This field of knowledge is commonly known as geography. Kant also divided geography into a number of sub-disciplines. He recognized the following six branches: Physical, mathematical, moral, political, commercial, and theological geography.

Geographic knowledge saw strong growth in Europe and the United States in the 1800s. This period also saw the emergence of a number of societies interested in geographic issues. In Germany, Alexander von Humboldt, Carl Ritter, and Fredrich Ratzel made substantial contributions to human and physical geography. Humboldt's publication Kosmos (1844) examines the geology and physical geography of the Earth. This work is considered by many academics to be a milestone contribution to geographic scholarship.

Late in the 19th Century, Ratzel theorized that the distribution and culture of the Earth's various human populations was strongly influenced by the natural environment. The French geographer Paul Vidal de la Blanche opposed this revolutionary idea. Instead, he suggested that human beings were a dominant force shaping the form of the environment. The idea that humans were modifying the physical environment was also prevalent in the United States.

In 1847, George Perkins Marsh gave an address to the Agricultural Society of Rutland County, Vermont. The subject of this speech was that human activity was having a destructive impact on land, especially through deforestation and land conversion. This speech also became the foundation for his book *Man and Nature* or *The Earth as Modified by Human Action*, first published in 1864. In this publication, Marsh warned of the ecological consequences of the continued development of the American frontier.

During the first 50 years of the 1900s, many academics in the field of geography extended the various ideas presented in the previous century to studies of small regions all over the world. Most of these studies used descriptive field methods to test research questions. Starting in about 1950, geographic research experienced a shift in methodology. Geographers began adopting a more scientific approach that relied on quantitative techniques. The quantitative revolution was also associated with a change in the way in which geographers studied the Earth and its phenomena. Researchers now began investigating process rather than mere description of the event of interest. Today, the quantitative approach is becoming even more prevalent due to advances in computer and software technologies.

In 1964, William Pattison published an article in the *Journal of Geography* (1964, 63: 211-216) that suggested that modern Geography was now composed of the following four academic traditions:

- Spatial Tradition - the investigation of the phenomena of geography from a strictly spatial perspective.
- Area Studies Tradition - the geographical study of an area on the Earth at either the local, regional, or global scale.
- Human-Land Tradition - the geographical study of human interactions with the environment.

Earth Science Tradition - the study of natural phenomena from a spatial perspective. This tradition is best described as theoretical physical geography.

Today, the academic traditions described by Pattison are still dominant fields of geographical investigation. However, the frequency and magnitude of human mediated environmental problems has been on a steady increase since the publication of this notion. These increases are the result of a growing human population and the consequent increase in the consumption of natural resources. As a result, an increasing number of researchers in geography are studying how humans modify the environment.

A significant number of these projects also develop strategies to reduce the negative impact of human activities on nature. Some of the dominant themes in these studies include: environmental degradation of the hydrosphere, atmosphere, lithosphere, and biosphere; resource use issues; natural hazards; environmental impact assessment; and the effect of urbanization and land-use change on natural environments.

Considering all of the statements presented concerning the history and development of geography, we are now ready to formulate a somewhat coherent definition. This definition suggests that geography, in its simplest form, is the field of knowledge that is concerned with how phenomena are spatially organized. Physical geography attempts to determine why natural phenomena have particular spatial patterns and orientation. This online textbook will focus primarily on the Earth Science Tradition. Some of the information that is covered in this textbook also deals with the alterations of the environment because of human interaction. These pieces of information belong in the Human-Land Tradition of geography.

Elements of Geography

In the previous, we discovered that geography consists of at least two different sub-fields of knowledge with similar methodology: Physical geography and human geography. The following table also helps to make the differences between these two types of geography more apparent. This table describes some of the phenomena or elements studied by each of these sub-fields of knowledge. Knowing what kinds of things are studied by geographers provides us with a better understanding of the differences between physical and human geography.

Table: *Some of the phenomena studied in physical and human geography.*

Physical Geography	Human Geography
Rocks and Minerals	Population
Landforms	Settlements
Soils	Economic Activities
Animals	Transportation
Plants	Recreational Activities
Water	Religion
Atmosphere	Political Systems
Rivers and Other Water Bodies	Social Traditions
Environment	Human Migration
Climate and Weather	Agricultural Systems
Oceans	Urban Systems

Geography is also a discipline that integrates a wide variety of subject matter. Almost any area of human knowledge can be examined from a spatial perspective. Figure describes some of the main subdisciplines within human and physical geography. Physical geography's primary subdisplines study the Earth's atmosphere (meteorology and climatology), animal and plant life (biogeography), physical landscape (geomorphology), soils (pedology), and waters (hydrology).

Some of the dominant areas of study in human geography include: human society and culture (social and cultural geography), behaviour (behavioural geography), economics (economic geography), politics (political geography), and urban systems (urban geography).

The graphic model indicates that the study of geography can also involve a holistic synthesis. Holistic synthesis connects knowledge from a variety of academic fields in both human and physical geography. For example, the study of the enhancement of the Earth's greenhouse effect and the resulting global warming requires a multidisciplinary approach for complete understanding.

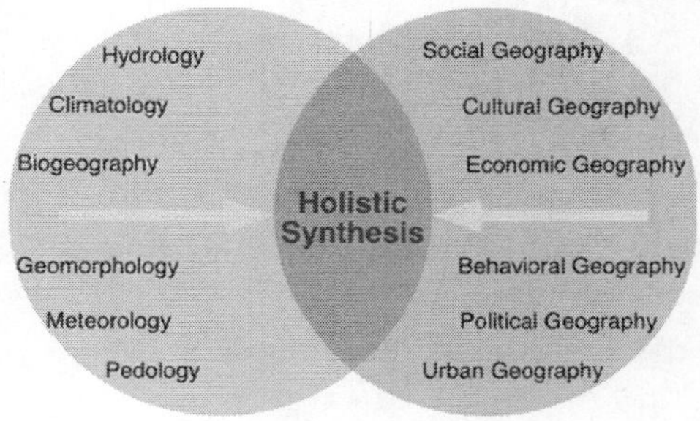

Figure: *Major subdisciplines of physical and human geography.*

The fields of climatology and meteorology are required to understand the physical effects of adding addition greenhouse gases to the atmosphere's radiation balance. The field of economic geography provides information on how various forms of human economic activity contribute to the emission of greenhouse gases through fossil fuel burning and land-use change. Combining the knowledge of both of these academic areas gives us a more comprehensive understanding of why this serious environmental problem occurs.

The holistic nature of geography is both a strength and a weakness. Geography's strength comes from its ability to connect functional interrelationships that are not normally noticed in narrowly defined fields of knowledge. The most obvious weakness associated with the geographical approach is related to the fact that holistic understanding is often too simple and misses important details of cause and effect.

Scope of Physical Geography

We have now learned that physical geography examines and investigates natural phenomena spatially. In the previous, we identified some of the key elements studied by physical geographers. Combining these two items, we can now suggest that physical geography *studies the spatial patterns of weather and climate, soils, vegetation, animals, water in all its forms, and landforms*. Physical geography also examines the interrelationships of these phenomena to human activities.

This sub-field of geography is academically known as the Human-Land Tradition. This area of geography has seen very keen interest and growth in the last few decades because of the acceleration of human induced environmental degradation. Thus, physical geography's scope is much broader than the simple spatial study of nature. It also involves the investigation of how humans are influencing nature.

Academics studying physical geography and other related earth sciences are rarely generalists. Most are in fact highly specialized in their fields of knowledge and tend to focus themselves in one of the following well defined areas of understanding in physical geography:

- Geomorphology - studies the various landforms on the Earth's surface.
- Pedology - is concerned with the study of soils.
- Biogeography - is the science that investigates the spatial relationships of plants and animals.
- Hydrology - is interested in the study of water in all its forms.
- Meteorology - studies the circulation of the atmosphere over short time spans.
- Climatology - studies the effects of weather on life and examines the circulation of the atmosphere over longer time spans.
- The above fields of knowledge generally have a primary role in introductory textbooks dealing with physical geography. Introductory physical geography textbooks can also contain information from other related disciplines including:
- Geology - studies the form of the Earth's surface and subsurface, and the processes that create and modify it.
- Ecology - the scientific study of the interactions between organisms and their environment.
- Oceanography - the science that examines the biology, chemistry, physics, and geology of oceans.
- Cartography - the technique of making maps.
- Astronomy - the science that examines celestial bodies and the cosmos.

History of Physical Geography

The nature of understanding in physical geography has changed over time. When investigating this change it becomes apparent that certain universal ideas or forces had very important ramifications to the academic study of physical geography. During the period from 1850 to 1950, there seems to be four main ideas that had a strong influence on the discipline:

- *Uniformitarianism:* This theory rejected the idea that catastrophic forces were responsible for the current conditions on the Earth. It suggested instead that continuing uniformity of existing processes were responsible for the present and past conditions of this planet.

- *Evolution:* Charles Darwin's Origin of Species (1859) suggested that natural selection determined which individuals would pass on their genetic traits to future generations. As a result of this theory, evolutionary explanations for a variety of natural phenomena were postulated by scientists. The theories of *uniformitarianism* and *evolution* arose from a fundamental change in the way humans explained the universe and nature. During the 16th, 17th, and 18th centuries scholars began refuting belief or myth based explanations of the cosmos, and instead used science to help explain the mysteries of nature. Belief based explanations of the cosmos are made consistent with a larger framework of knowledge that focuses on some myth. However, theories based on science questioned the accuracy of these beliefs.
- *Exploration and Survey:* Much of the world had not been explored before 1900. Thus, during this period all of the fields of physical geography were actively involved with basic data collection. This data collection included activities like determining the elevation of land surfaces, classification and description of landforms, the measurement of the volume of flow of rivers, measurement of various phenomena associated to weather and climate, and the classification of soils, organisms, biological communities and ecosystems.
- *Conservation:* Beginning in the 1850s a concern for the environment began to develop as a result of the human development of once natural areas in the United States and Europe. One of the earliest statements of these ideas came from George Perkins Marsh (1864) in his book "*Man in Nature*" or "*Physical Geography as Modified by Human Action*". This book is often cited by scholars as the first significant academic contribution to conservation and environmentalism.

After 1950, the following two forces largely determined the nature of physical geography:

- *The Quantitative Revolution:* Measurement became the central focus of research in Physical Geography. It was used primarily for hypothesis testing. With measurement came mapping, models, statistics, mathematics, and hypothesis testing. The quantitative revolution was also associated with a change in the way in which physical geographers studied the Earth and its phenomena. Researchers now began investigating process rather than mere description of the environment.

- *The study of Human/Land Relationships:* The influence of human activity on the environment was becoming very apparent after 1950. As a result, many researchers in physical geography began studying the influence of humans on the environment. Some of the dominant themes in these studies included: environmental degradation and resource use; natural hazards and impact assessment; and the effect of urbanization and land-use change on natural environments.

Future of Physical Geography

The following list describes some of the important future trends in physical geography research:

- Continued development of applied physical geography for the analysis and correction of human-induced environmental problems. A student of applied physical geography uses theoretical information from the field of physical geography to manage and solve problems related to natural phenomena found in the real world.
- Remote Sensing - Advances in technology have caused the development of many new instruments for the monitoring of the Earth's resources and environment from airborne and space platforms. The most familiar use of remote sensing technology is to monitor the Earth's weather for forecasting.

Sub-Fields Of Physical Geography

Geomorphology

The surface of Earth is modified by a combination of surface processes that sculpt landscapes, and geologic processes that cause tectonic uplift and subsidence. Surface processes comprise the action of water, wind, ice, fire, and living things on the surface of the Earth, along with chemical reactions that form soils and alter material properties, the stability and rate of change of topography under the force of gravity, and other factors, such as (in the very recent past) human alteration of the landscape.

Many of these factors are strongly mediated by climate. Geologic processes include the uplift of mountain ranges, the growth of volcanoes, isostatic changes in land surface elevation (sometimes in response to surface processes), and the formation of deep sedimentary basins where the surface of Earth drops and is filled with material eroded from other parts of the landscape. The Earth surface and its topography therefore are an intersection of climatic, hydrologic, and biologic action with geologic processes.

The broad-scale topographies of Earth illustrate this intersection of surface and subsurface action. Mountain belts are uplifted due to geologic processes. Denudation of these high uplifted regions produces sediment that is transported and deposited elsewhere within the landscape or off the coast. On progressively smaller scales, similar ideas apply, where individual landforms evolve in response to the balance of additive processes (uplift and deposition) and subtractive processes (subsidence and erosion).

Often, these processes directly affect each other: ice sheets, water, and sediment are all loads that change topography through flexural isostasy. Topography can modify the local climate, for example through orographic precipitation, which in turn modifies the topography by changing the hydrologic regime in which it evolves. Many geomorphologists are particularly interested in the potential for feedbacks between climate and tectonics mediated by geomorphic processes.

In addition to these broad-scale questions, geomorphologists address issues that are more specific and/or more local. Glacial geomorphologists investigate glacial deposits such as moraines, eskers, and proglacial lakes, as well as glacial erosional features, to build chronologies of both small glaciers and large ice sheets and understand their motions and effects upon the landscape.

Fluvial geomorphologists focus on rivers, how they transport sediment, migrate across the landscape, cut into bedrock, respond to environmental and tectonic changes, and interact with humans. Soils geomorphologists investigate soil profiles and chemistry to learn about the history of a particular landscape and understand how climate, biota, and rock interact. Other geomorphologists study how hillslopes form and change. Still others investigate the relationships between ecology and geomorphology. Because geomorphology is defined to comprise everything related to the surface of Earth and its modification, it is a broad field with many facets. Practical applications of geomorphology include hazard assessment (such as landslide prediction and mitigation), river control and stream restoration, and coastal protection.

History

With some notable exceptions, geomorphology is a relatively young science, growing along with interest in other aspects of the earth sciences in the mid-19th century.

Ancient Geomorphology

The first theory of geomorphology was arguably devised by the polymath Chinese scientist and statesman Shen Kuo (1031-1095 AD). This was based on his observation of marine fossil shells in a geological stratum of a mountain hundreds of miles from the Pacific Ocean. Noticing bivalve shells running in a horizontal span along the cut section of a cliffside, he theorized that the cliff was once the pre-historic location of a seashore that had shifted hundreds of miles over the centuries.

He inferred that the land was reshaped and formed by soil erosion of the mountains and by deposition of silt, after observing strange natural erosions of the Taihang Mountains and the Yandang Mountain near Wenzhou. Furthermore, he promoted the theory of gradual climate change over centuries of time once ancient petrified bamboos were found to be preserved underground in the dry, northern climate zone of *Yanzhou*, which is now modern day Yan'an, Shaanxi province.

Early Modern Geomorphology

The first use of the word geomorphology was likely to be in the German language when it appeared in Laumann's 1858 work. Keith Tinkler has suggested that the word came into general use in English, German and French after John Wesley Powell and W. J. McGee used it in the International Geological Conference of 1891. An early popular geomorphic model was the *geographical cycle* or the *cycle of erosion*, developed by William Morris Davis between 1884 and 1899. The cycle was inspired by theories of uniformitarianism first formulated by James Hutton (1726–1797). Concerning valley forms, uniformitarianism depicted the cycle as a sequence in which a river cuts a valley more and more deeply, but then erosion of side valleys eventually flatten the terrain again, to a lower elevation.

Tectonic uplift could start the cycle over. Many studies in geomorphology in the decades following Davis' development of his theories sought to fit their ideas into this framework for broad scale landscape evolution, and are often today termed "Davisian". Davis' ideas have largely been superseded today, mainly due to their lack of predictive power and qualitative nature, but he remains an extremely important figure in the history of the subject. In the 1920s, Walther Penck developed an alternative model to Davis', believing that landform evolution was better described as a balance between ongoing processes of uplift and denudation, rather than Davis' single uplift followed by decay. However, due to his relatively young death, disputes with Davis and a lack of English translation of his work his ideas were not widely recognised for many years.

These authors were both attempting to place the study of the evolution of the Earth's surface on a more generalized, globally relevant footing than had existed before. In the earlier parts of the 19th century, authors - especially in Europe - had tended to attribute the form of landscape to local climate, and in particular to the specific effects of glaciation and periglacial processes. In contrast, both Davis and Penck were seeking to emphasize the importance of evolution of landscapes through time and the generality of Earth surface processes across different landscapes under different conditions.

Quantitative Geomorphology

While Penck and Davis and their followers were writing and studying primarily in Western Europe, another, largely separate, school of geomorphology was developed in the United States in the middle years of the 20th century. Following the early trailblazing work of Grove Karl Gilbert around the turn of the 20th century, a group of natural scientists, geologists and hydraulic engineers including Ralph Alger Bagnold, John Hack, Luna Leopold, Thomas Maddock and Arthur Strahler began to research the form of landscape elements such as rivers and hillslopes by taking systematic, direct, quantitative measurements of aspects of them and investigating the scaling of these measurements. These methods began to allow prediction of the past and future behaviour of landscapes from present observations, and were later to develop into what the modern trend of a highly quantitative approach to geomorphic problems.

Quantitative geomorphology can involve fluid dynamics and solid mechanics, geomorphometry, laboratory studies, field measurements, theoretical work, and full landscape evolution modelling. These approaches are used to understand weathering and the formation of soils, sediment transport, landscape change, and the interactions between climate, tectonics, erosion, and deposition.

Contemporary Geomorphology

Today, the field of geomorphology encompasses a very wide range of different approaches and interests. Modern researchers aim to draw out quantitative "laws" that govern Earth surface processes, but equally, recognize the uniqueness of each landscape and environment in which these processes operate. Particularly important realizations in contemporary geomorphology include:

- that not all landscapes can be considered as either "stable" or "perturbed", where this perturbed state is a temporary displacement away from some ideal target form. Instead, dynamic changes of the landscape are now seen as an essential part of their nature.
- that many geomorphic systems are best understood in terms of the stochasticity of the processes occurring in them, that is, the probability distributions of event magnitudes and return times. This in turn has indicated the importance of chaotic determinism to landscapes, and that landscape properties are best considered statistically. The same processes in the same landscapes does not always lead to the same end results.

Processes

Modern geomorphology focuses on the quantitative analysis of interconnected processes. Modern advances in geochronology, in particular cosmogenic radionuclide dating, optically stimulated luminescence dating and low-temperature thermochronology have enabled us for the first time to measure the rates at which geomorphic processes occur on geological timescales. At the same time, the use of more precise physical measurement techniques, including differential GPS, remotely sensed digital terrain models and laser scanning techniques, have allowed quantification and study of these processes as they happen. Computer simulation and modelling may then be used to test our understanding of how these processes work together and through time.

Geomorphically relevant processes generally fall into (1) the production of regolith by weathering and erosion, (2) the transport of that material, and (3) its eventual deposition. Although there is a general movement of material from uplands to lowlands, erosion, transport, and deposition often occur in closely spaced tandem all across the landscape. The nature of the processes investigated by geomorphologists is strongly dependent on the landscape or landform under investigation and the time and length scales of interest. However, the following non-exhaustive list provides a flavour of the landscape elements associated with some of these. Primary surface processes responsible for most topographic features include wind, waves, chemical dissolution, mass wasting, groundwater movement, surface water flow, glacial action, tectonism, and volcanism. Other more exotic geomorphic processes might include periglacial (freeze-thaw) processes, salt-mediated action, or extraterrestrial impact.

Fluvial Processes

Rivers and streams are not only conduits of water, but also of sediment. The water, as it flows over the channel bed, is able to mobilize sediment and transport it downstream, either as bed load, suspended load or dissolved load. The rate of sediment transport depends on the availability of sediment itself and on the river's discharge. Rivers are also capable of eroding into rock and creating new sediment, both from their own beds and also by coupling to the surrounding hillslopes. In this way, rivers are thought of as setting the base level for large scale landscape evolution in nonglacial environments. Rivers are key links in the connectivity of different landscape elements. As rivers flow across the landscape, they generally increase in size, merging with other rivers. The network of rivers thus formed is a drainage system and is often dendritic (tree-like), but may adopt other patterns depending on the regional topography and underlying geology.

Eolian Processes

Eolian processes pertain to the activity of the winds and more specifically, to the winds' ability to shape the surface of the Earth. Winds may erode, transport, and deposit materials, and are effective agents in regions with sparse vegetation and a large supply of fine, unconsolidated sediments. Although water and mass flow tend to mobilize more material than wind in most environments, eolian processes are important in arid environments such as deserts.

Glacial Processes

Glaciers, while geographically restricted, are effective agents of landscape change. The gradual movement of ice down a valley causes abrasion and plucking of the underlying rock. Abrasion produces fine sediment, termed glacial flour. The debris transported by the glacier, when the glacier recedes, is termed a moraine. Glacial erosion is responsible for U-shaped valleys, as opposed to the V-shaped valleys of fluvial origin. The way glacial processes interact with other landscape elements, particularly hillslope and fluvial processes, is an important aspect of Plio-Pleistocene landscape evolution and its sedimentary record in many high mountain environments.

Environments that have been relatively recently glaciated but are no longer may still show elevated landscape change rates compared to those that have never been glaciated. Nonglacial geomorphic processes which nevertheless have been conditioned by past glaciation are termed paraglacial processes. This concept contrasts with periglacial processes, which are directly driven by formation or melting of ice or frost.

Tectonic Processes

Tectonic effects on geomorphology can range from scales of millions of years to minutes or less. The effects of tectonics on landscape are heavily dependent on the nature of the underlying bedrock fabric that more less controls what kind of local morphology tectonics can shape. Earthquakes can, in terms of minutes, submerge large areas of land creating new wetlands. Isostatic rebound can account for significant changes over thousand or hundreds of years, and allows erosion of a mountain belt to promote further erosion as mass is removed from the chain and the belt uplifts. Long-term plate tectonic dynamics give rise to orogenic belts, large mountain chains with typical lifetimes of many tens of millions of years, which form focal points for high rates of fluvial and hillslope processes and thus long-term sediment production. Features of deeper mantle dynamics such as plumes and delamination of the lower lithosphere have also been hypothesised to play important roles in the long term (> million year), large scale (thousands of km) evolution of the Earth's topography. Both can promote surface uplift through isostasy as hotter, less dense, mantle rocks displace cooler, denser, mantle rocks at depth in the Earth.

Biological Processes

The interaction of living organisms with landforms, or biogeomorphologic processes, can be of many different forms, and is probably of profound importance for the terrestrial geomorphic system as a whole. Biology can influence very many geomorphic processes, ranging from biogeochemical processes controlling chemical weathering, to the influence of mechanical processes like burrowing and tree throw on soil development, to even controlling global erosion rates through modulation of climate through carbon dioxide balance. Terrestrial landscapes in which the role of biology in mediating surface processes can be definitively excluded are extremely rare, but may hold important information for understanding the geomorphology of other planets, such as Mars.

Overlap with Other Fields

There is a considerable overlap between geomorphology and other fields. Deposition of material is extremely important in sedimentology. Weathering is the chemical and physical disruption of earth materials in place on exposure to atmospheric or near surface agents, and is typically studied by soil scientists and environmental chemists, but is an essential component of geomorphology because it is what provides the material that can be moved in the first place. Civil and environmental engineers are concerned with erosion and sediment transport, especially related to canals, slope stability (and natural hazards), water quality, coastal environmental management, transport of contaminants, and stream restoration. Glaciers can cause extensive erosion and deposition in a short period of time, making them extremely important entities in the high latitudes and meaning that they set the conditions in the headwaters of mountain-born streams; glaciology therefore is important in geomorphology.

Hydrology

Hydrology is the study of the movement, distribution, and quality of water on Earth and other planets, including the hydrologic cycle, water resources and environmental watershed sustainability. A practitioner of hydrology is a hydrologist, working within the fields of earth or environmental science, physical geography, geology or civil and environmental engineering. Domains of hydrology include hydrometeorology, surface hydrology, hydrogeology, drainage basin management and water quality, where water plays the central role. Oceanography and meteorology are not included because water is only one of many important aspects within those fields. Hydrological research can inform environmental engineering, policy and planning.

History

Hydrology has been a subject of investigation and engineering for millennia. For example, about 4000 B.C. the Nile was dammed to improve agricultural productivity of previously barren lands. Mesopotamian towns were protected from flooding with high earthen walls. Aqueducts were built by the Greeks and Ancient Romans, while the history of China shows they built irrigation and flood control works.

The ancient Sinhalese used hydrology to build complex irrigation works in Sri Lanka, also known for invention of the Valve Pit which allowed construction of large reservoirs, anicuts and canals which still function. Marcus Vitruvius, in the first century B.C., described a philosophical theory of the hydrologic cycle, in which precipitation falling in the mountains infiltrated the Earth's surface and led to streams and springs in the lowlands. With adoption of a more scientific approach, Leonardo da Vinci and Bernard Palissy independently reached an accurate representation of the hydrologic cycle. It was not until the 17th century that hydrologic variables began to be quantified.

Pioneers of the modern science of hydrology include Pierre Perrault, Edme Mariotte and Edmund Halley. By measuring rainfall, runoff, and drainage area, Perrault showed that rainfall was sufficient to account for flow of the Seine. Marriotte combined velocity and river cross-section measurements to obtain discharge, again in the Seine. Halley showed that the evapouration from the Mediterranean Sea was sufficient to account for the outflow of rivers flowing into the sea. Advances in the 18th century included the Bernoulli piezometre and Bernoulli's equation, by Daniel Bernoulli, the Pitot tube. The 19th century saw development in groundwater hydrology, including Darcy's law, the Dupuit-Thiem well formula, and Hagen-Poiseuille's capillary flow equation. Rational analyses began to replace empiricism in the 20th century, while governmental agencies began their own hydrological research programmes. Of particular importance were Leroy Sherman's unit hydrograph, the infiltration theory of Robert E. Horton, and C.V. Theis's aquifer test/equation describing well hydraulics. Since the 1950s, hydrology has been approached with a more theoretical basis than in the past, facilitated by advances in the physical understanding of hydrological processes and by the advent of computers and especially geographic information systems (GIS).

Uncertainty Analyses

Remote Sensing

Remote sensing of hydrologic processescan provide information of various types.. Sources include land based sensors, airborne sensors and Satellite sensors. Information can include clouds, surface moisture, vegetation cover.

Water Quality

In hydrology, studies of water quality concern organic and inorganic compounds, and both dissolved and sediment material. In addition, water quality is affected by the interaction of dissolved oxygen with organic material and various chemical transformations that may take place. Measurements of water quality may involve either in-situ methods, in which analyses take place on-site, often automatically, and laboratory-based analyses and may include microbiological analysis.

Integrating Measurement and Modelling

- Budget analyses
- Parametre estimation
- Scaling in time and space
- Data assimilation
- Quality control of data

Prediction

Observations of hydrologic processes are used to make predictions of the future behaviour of hydrologic systems (water flow, water quality). One of the major current concerns in hydrologic research is "Prediction in Ungauged Basins" (PUB), i.e. in basins where no or only very few data exist.

Statistical Hydrology

By analysing the statistical properties of hydrologic records, such as rainfall or river flow, hydrologists can estimate future hydrologic phenomena. When making assessments of how often relatively rare events will occur, analyses are made in terms of the return period of such events. Other quantities of interest include the average flow in a river, in a year or by season. These estimates are important for engineers and economists so that proper risk analysis can be performed to influence investment decisions in future infrastructure and to determine the yield reliability characteristics of water supply systems. Statistical information is utilised to formulate operating rules for large dams forming part of systems which include agricultural, industrial and residential demands.

Modelling

Hydrological models are simplified, conceptual representations of a part of the hydrologic cycle. They are primarily used for hydrological prediction and for understanding hydrological processes. Two major types of hydrological models can be distinguished:

- Models based on data. These models are black box systems, using mathematical and statistical concepts to link a certain input (for instance rainfall) to the model output (for instance runoff). Commonly used techniques are regression, transfer functions, and system identification. The simplest of these models may be linear models, but it is common to deploy non-linear components to represent some general aspects of a catchment's response without going deeply into the real physical processes involved. An example of such an aspect is the well-known behaviour that a catchment will respond much more quickly and strongly when it is already wet than when it is dry..

- Models based on process descriptions. These models try to represent the physical processes observed in the real world. Typically, such models contain representations of surface runoff, subsurface flow, evapotranspiration, and channel flow, but they can be far more complicated. These models are known as deterministic hydrology models. Deterministic hydrology models can be subdivided into single-event models and continuous simulation models.

Recent research in hydrological modelling tries to have a more global approach to the understanding of the behaviour of hydrologic systems to make better predictions and to face the major challenges in water resources management.

Transport

Water movement is a significant means by which other material, such as soil or pollutants, are transported from place to place. Initial input to receiving waters may arise from a point source discharge or a line source or area source, such as surface runoff. Since the 1960s rather complex mathematical models have been developed, facilitated by the availability of high speed computers. The most common pollutant classes analyzed are nutrients, pesticides, total dissolved solids and sediment.

Biogeography

Biogeography is the study of the distribution of species (biology), organisms, and ecosystems in geographic space and through geological time. Organisms and biological communities vary in a highly regular fashion along geographic gradients of latitude, elevation, isolation and habitat area. Knowledge of spatial variation in the numbers and types of organisms is as vital to us today as it was to our early human ancestors, as we adapt to heterogeneous but geographically predictable environments. Biogeography is an integrative field of enquiry that unites concepts and information from ecology, evolutionary biology, geology, and physical geography. Modern biogeographic research combines information and ideas from many fields, from the physiological and ecological constraints on organismal dispersal to geological and climatological phenomena operating at global spatial scales and evolutionary time frames.

The patterns of species distribution across geographical areas can usually be explained through a combination of historical factors such as: speciation; extinction; continental drift; glaciation, and associated variations in sea level, river routes, and habitat; and river capture; in combination with the geographic constraints of landmass areas and isolation; and the available ecosystem energy supplies. Over periods of ecological changes, biogeography includes the study of plant and animal species in: their past and/or present living *refugium* habitat; their interim living sites; and/or their survival locales. As writer David Quammen put it, "...biogeography does more than ask *Which species?* and *Where*. It also asks *Why?* and, what is sometimes more crucial, *Why not?*."

Modern biogeography often employs the use of Geographic Information Systems (GIS), to understand the factors affecting organism distribution, and to predict future trends in organism distribution. Often mathematical models and GIS are employed to solve ecological problems that have a spatial aspect to them. Biogeography is most keenly observed on the world's islands. These habitats are often much more manageable areas of study because they are more condensed than larger ecosystems on the mainland. Islands are also ideal locations because they allow scientists to look at habitats that new species have only recently colonized and can observe how they disperse throughout the island, the success they achieve in these places, and they can then apply this information to similar mainland habitats.

Islands are very diverse in their biomes, ranging from the tropical to arctic climates. This diversity in habitat allows for a wider range of species study in different parts of the world. One scientist who recognized the importance of these geographic locations was Charles Darwin, who remarked in his journal "The Zoology of Archipelagoes will be well worth examination". Two chapters in *On the Origin of Species* were devoted to geographical distribution.

History

The scientific theory of biogeography grows out of the work of

Alexander von Humboldt (1769–1859), Hewett Cottrell Watson (1804–1881), Alphonse de Candolle (1806–1893), Alfred Russel Wallace (1823–1913), Philip Lutley Sclater (1829–1913) and other biologists and explorers. Wallace studied the distribution of flora and fauna in the Amazon Basin and the Malay Archipelago in the mid-19th century. Wallace and Sclater saw biogeography as a source of support for the theory of evolution. Key findings, such as the sharp difference in fauna either side of the Wallace Line, and the sharp difference that existed between North and South America prior to their relatively recent faunal interchange, can only be understood in this light. Otherwise, the field of biogeography would be seen as a purely descriptive one.

The publication of *The Theory of Island Biogeography* by Robert MacArthur and E.O. Wilson in 1967 showed that the species richness of an area could be predicted in terms of such factors as habitat area, immigration rate and extinction rate. This added to the long-standing interest in island biogeography. The application of island biogeography theory to habitat fragments spurred the development of the fields of conservation biology and landscape ecology.

Classic biogeography has been expanded by the development of molecular systematics, creating a new discipline known as phylogeography. This development allowed scientists to test theories about the origin and dispersal of populations, such as island endemics. For example, while classic biogeographers were able to speculate about the origins of species in the Hawaiian Islands, phylogeography allows them to test theories of relatedness between these populations and putative source populations in Asia and North America.

Paleobiogeography

Paleobiogeography goes one step further to include paleogeographic data and considerations of plate tectonics. Using molecular analyses and corroborated by fossils, it has been possible to demonstrate that perching birds evolved first in the region of Australia or the adjacent Antarctic (which at that time lay somewhat further north and had a temperate climate). From there, they spread to the other Gondwanan continents and Southeast Asia - the part of Laurasia then closest to their origin of dispersal – in the late Paleogene, before achieving a global distribution in the early Neogene.

Not knowing the fact that at the time of dispersal, the Indian Ocean was much narrower than it is today, and that South America was closer to the Antarctic, one would be hard pressed to explain the presence of many "ancient" lineages of perching birds in Africa, as well as the mainly South American distribution of the suboscines.

Paleobiogeography also helps constrain hypotheses on the timing of biogeographic events such as vicariance and geodispersal, and provides unique information on the formation of regional biotas. For example, data from species-level phylogenetic and biogeographic studies tell us that the Amazonian fish fauna accumulated incrementally over a period of tens of millions of years, principally by means of allopatric speciation, and in an arena extending over most of the area of tropical South America (Albert & Reis 2011). In other words, unlike some of the well-known insular faunas (Galapagos finches, Hawaiian drosophilid flies, African rift lake cichlids), the species-rich Amazonian ichthyofauna is not the result of recent adaptive radiations.

For freshwater organisms, landscapes are divided naturally into discrete drainage basins by watersheds, episodically isolated and reunited by erosional processes. In regions like the Amazon Basin with an exceptionally low (flat) topographic relief, the many waterways have had a highly reticulated history over geological time. In such a context stream capture is an important factor affecting the evolution and distribution of freshwater organisms. Stream capture occurs when an upstream portion of one river drainage is diverted to the downstream portion of an adjacent basin. This can happen because of tectonic uplift (or subsidence), natural damming as a result of a landslide, or by headward or lateral erosion of the watershed between adjacent basins.

Classification

Biogeography is a synthetic science, related to geography, biology, soil science, geology, climatology, ecology and evolution.

Some fundamental concepts in biogeography include:

- evolution – change in genetic composition of a population
- extinction – disappearance of a species
- dispersal – movement of populations away from their point of origin, related to migration
- geodispersal – the erosion of barriers to biotic dispersal and gene flow, that permit range expansion and the merging of previously isolated biotas
- range and distribution
- endemic areas

- vicariance – the formation of barriers to biotic dispersal and gene flow, that tend to subdivide species and biotas, leading to speciation and extinction

Climatology

Climatology is the study of climate, scientifically defined as weather conditions averaged over a period of time, and is a branch of the atmospheric sciences. Basic knowledge of climate can be used within shorter term weather forecasting using analog techniques such as the El Niño – Southern Oscillation (ENSO), the Madden-Julian Oscillation (MJO), the North Atlantic Oscillation (NAO), the Northern Annualar Mode (NAM), the Arctic oscillation (AO), the Northern Pacific (NP) Index, the Pacific Decadal Oscillation (PDO), and the Interdecadal Pacific Oscillation (IPO). Climate models are used for a variety of purposes from study of the dynamics of the weather and climate system to projections of future climate.

History

The earliest person to hypothesize climate change may have been the medieval Chinese scientist Shen Kuo (1031–1095). Shen Kuo theorized that climates naturally shifted over an enormous span of time, after observing petrified bamboos found underground near Yanzhou (modern day Yan'an, Shaanxi province), a dry-climate area unsuitable for the growth of bamboo. Early climate researchers include Edmund Halley, who published a map of the trade winds in 1686 after a voyage to the southern hemisphere. Benjamin Franklin (1706-1790) first mapped the course of the Gulf Stream for use in sending mail from the United States to Europe. Francis Galton (1822-1911) invented the term *anticyclone*. Helmut Landsberg (1906-1985) fostered the use of statistical analysis in climatology, which led to its evolution into a physical science.

Different Approaches

Climatology is approached in a variety of ways. Paleoclimatology seeks to reconstruct past climates by examining records such as ice cores and tree rings (dendroclimatology). Paleotempestology uses these same records to help determine hurricane frequency over millennia. The study of contemporary climates incorporates meteorological data accumulated over many years, such as records of rainfall, temperature and atmospheric composition. Knowledge of the atmosphere and its dynamics is also embodied in models, either statistical or mathematical, which help by integrating different observations and testing how they fit together. Modelling is used for understanding past, present and potential future climates. Historical climatology is the study of climate as related to human history and thus focuses only on the last few thousand years. Climate research is made difficult by the large scale, long time periods, and complex processes which govern climate. Climate is governed by physical laws which can be expressed as differential equations. These equations are coupled and nonlinear, so that approximate solutions are obtained by using numerical methods to create global climate models. Climate is sometimes modelled as a stochastic process but this is generally accepted as an approximation to processes that are otherwise too complicated to analyze.

Madden–Julian Oscillation

The Madden–Julian Oscillation (MJO) is an equatorial travelling pattern of anomalous rainfall that is planetary in scale. It is characterized by an eastward progression of large regions of both enhanced and suppressed tropical rainfall, observed mainly over the Indian and Pacific Oceans. The anomalous rainfall is usually first evident over the western Indian Ocean, and remains evident as it propagates over the very warm ocean waters of the western and central tropical Pacific.

This pattern of tropical rainfall then generally becomes very nondescript as it moves over the cooler ocean waters of the eastern Pacific but reappears over the tropical Atlantic and Indian Oceans. The wet phase of enhanced convection and precipitation is followed by a dry phase where convection is suppressed. Each cycle lasts approximately 30–60 days. The MJO is also known as the 30–60 day oscillation, 30–60 day wave, or intraseasonal oscillation.

Northern Annular Mode (NAM) or Arctic Oscillation (AO)

The NAM, or AO, is defined as the first EOF of northern hemisphere winter SLP data from the tropics and subtropics. It explains 23% of the average winter (December–March) variance, and it is dominated by the NAO structure in the Atlantic. Although there are some subtle differences from the regional pattern over the Atlantic and Arctic, the main difference is larger amplitude anomalies over the North Pacific of the same sign as those over the Atlantic. This feature gives the NAM a more annular (or zonally symmetric) structure.

Interdecadal Pacific Oscillation (IPO)

The Interdecadal Pacific Oscillation (IPO or ID) display similar sea surface temperature (SST) and sea level pressure patterns to the

PDO, with a cycle of 15–30 years, but affects both the north and south Pacific. In the tropical Pacific, maximum SST anomalies are found away from the equator. This is quite different from the quasi-decadal oscillation (QDO) with a period of 8–12 years and maximum SST anomalies straddling the equator, thus resembling ENSO.

Models

Climate models use quantitative methods to simulate the interactions of the atmosphere, oceans, land surface, and ice. They are used for a variety of purposes from study of the dynamics of the weather and climate system to projections of future climate. All climate models balance, or very nearly balance, incoming energy as short wave (including visible) electromagnetic radiation to the earth with outgoing energy as long wave (infrared) electromagnetic radiation from the earth. Any unbalance results in a change in the average temperature of the earth.

The most talked-about models of recent years have been those relating temperature to emissions of carbon dioxide. These models predict an upward trend in the surface temperature record, as well as a more rapid increase in temperature at higher latitudes.

Models can range from relatively simple to quite complex:

- A simple radiant heat transfer model that treats the earth as a single point and averages outgoing energy
- this can be expanded vertically (radiative-convective models), or horizontally
- finally, (coupled) atmosphere–ocean–sea ice global climate models discretise and solve the full equations for mass and energy transfer and radiant exchange.

Differences with Meteorology

In contrast to meteorology, which focuses on short term weather systems lasting up to a few weeks, climatology studies the frequency and trends of those systems. It studies the periodicity of weather events over years to millennia, as well as changes in long-term average weather patterns, in relation to atmospheric conditions. Climatologists, those who practice climatology, study both the nature of climates – local, regional or global – and the natural or human-induced factors that cause climates to change. Climatology considers the past and can help predict future climate change. Phenomena of climatological interest include the atmospheric boundary layer, circulation patterns, heat transfer (radiative, convective and latent), interactions between the atmosphere and the oceans and land surface (particularly vegetation, land use and topography), and the chemical and physical composition of the atmosphere.

Use in Weather Forecasting

A more complicated way of making a forecast, the analog technique requires remembering a previous weather event which is expected to be mimicked by an upcoming event. What makes it a difficult technique to use is that there is rarely a perfect analog for an event in the future. Some call this type of forecasting pattern recognition, which remains a useful method of observing rainfall over data voids such as oceans with knowledge of how satellite imagery relates to precipitation rates over land, as well as the forecasting of precipitation amounts and distribution in the future. A variation on this theme is used in Medium Range forecasting, which is known as teleconnections, when you use systems in other locations to help pin down the location of another system within the surrounding regime. One method of using teleconnections are by using climate indices such as ENSO-related phenomena.

Meteorology

Meteorology is the interdisciplinary scientific study of the atmosphere. Studies in the field stretch back millennia, though significant progress in meteorology did not occur until the 18th century. The 19th century saw breakthroughs occur after observing networks developed across several countries. After the development of the computer in the latter half of the 20th century, breakthroughs in weather forecasting were achieved. Meteorological phenomena are observable weather events which illuminate, and are explained by the science of meteorology. Those events are bound by the variables that exist in Earth's atmosphere; temperature, air pressure, water vapour, and the gradients and interactions of each variable, and how they change in time. Different spatial scales are studied to determine how systems on local, regional, and global levels impact weather and climatology. Meteorology, climatology, atmospheric physics, and atmospheric chemistry are sub-disciplines of the atmospheric sciences. Meteorology and hydrology compose the interdisciplinary field of hydrometeorology. Interactions between Earth's atmosphere and the oceans are part of coupled ocean-atmosphere studies. Meteorology has application in many diverse fields such as the military, energy production, transport, agriculture and construction.

Timeline of Meteorology

The Timeline of Meteorology contains events of scientific and technological advancements in the area of atmospheric sciences. The most notable advancements in observational meteorology, weather forecasting, climatology, atmospheric chemistry, and atmospheric physics are listed chronologically. Some historical weather events are included that mark time periods where advancements were made, or even that sparked policy change.

Antiquity

- 3000 BC - The beginnings of meteorology can be traced back in India to 3000 B.C.E, such as the Upanishads, contain serious discussion about the processes of cloud formation and rain and the seasonal cycles caused by the movement of earth round the sun.
- 600 BC - Thales may qualify as the first Greek meteorologist. He described the water cycle in a fairly accurate way. He also issued the first seasonal crop forecast.
- 400 BC - There is some evidence that Democritus predicted changes in the weather, and that he used this ability to convince people that he could predict other future events.
- 400 BC - Hippocrates writes a treatise called *Airs, Waters and Places*, the earliest known work to include a discussion of weather. More generally, he wrote about common diseases that occur in particular locations, seasons, winds and air.
- 350 BC - Aristotle writes *Meteorology*.

Although the term meteorology is used today to describe a subdiscipline of the atmospheric sciences, Aristotle's work is more general. The work touches upon much of what is known as the earth sciences. In his own words:

> *...all the affections we may call common to air and water, and the kinds and parts of the earth and the affections of its parts.*

One of the most impressive achievements in *Meteorology* is his description of what is now known as the hydrologic cycle:

> *Now the sun, moving as it does, sets up processes of change and becoming and decay, and by its agency the finest and sweetest water is every day carried up and is dissolved into vapour and rises to the upper region,*

where it is condensed again by the cold and so returns to the earth.

- 250 BC - Archimedes studies the concepts of buoyancy and the hydrostatic principle. Positive buoyancy is necessary for the formation of convective clouds (cumulus, cumulus congestus and cumulonimbus).
- 25 AD - Pomponius Mela, a geographer for the Roman empire, formalizes the climatic zone system.
- c. 80 AD - In his *Lunheng*, the Han Dynasty Chinese philosopher Wang Chong (27-97 AD) dispels the Chinese myth of rain coming from the heavens, and states that rain is evapourated from water on the earth into the air and forms clouds, stating that clouds condense into rain and also form dew, and says when the clothes of people in high mountains are moistened, this is because of the air-suspended rain water. However, Wang Chong supports his theory by quoting a similar one of Gongyang Gao's, the latter's commentary on the *Spring and Autumn Annals*, the Gongyang Zhuan, compiled in the 2nd century BC, showing that the Chinese conception of rain evapourating and rising to form clouds goes back much farther than Wang Chong. Wang Chong wrote:

As to this coming of rain from the mountains, some hold that the clouds carry the rain with them, dispersing as it is precipitated (and they are right). Clouds and rain are really the same thing. Water evapourating upwards becomes clouds, which condense into rain, or still further into dew.

Middle Ages

- 500 AD - Varahamithra's classical work, Brihatsamhita, written about 500 A.D., provides a clear evidence that a deep knowledge of atmospheric processes existed in India even in those times.
- 7th century-Kalidasa in his epic Meghaduta mentions the date of onset of Monsoon over central India and traces the path of monsoon clouds,.
- 7th century-St. Isidore of Seville,in his work *De Rerum Natura*, writes about astronomy, cosmology and meteorology. In the chapter dedicated to Meteorology, he discusses the thunder, clouds, rainbows and wind.
- 9th century - Al-Kindi (Alkindus), an Arab naturalist, writes a treatise on meteorology entitled *Risala fi l-Illa al-Failali l-*

Madd wa l-Fazr (*Treatise on the Efficient Cause of the Flow and Ebb*), in which he presents an argument on tides which "depends on the changes which take place in bodies owing to the rise and fall of temperature."

- 9th century - Al-Dinawari, a Kurdish naturalist, writes the *Kitab al-Nabat* (*Book of Plants*), in which he deals with the application of meteorology to agriculture during the Muslim Agricultural Revolution. He describes the meteorological character of the sky, the planets and constellations, the Sun and Moon, the lunar phases indicating seasons and rain, the *anwa* (heavenly bodies of rain), and atmospheric phenomena such as winds, thunder, lightning, snow, floods, valleys, rivers, lakes, wells and other sources of water.
- 10th century - Ibn Wahshiyya's *Nabatean Agriculture* discusses the weather forecasting of atmospheric changes and signs from the planetary astral alterations; signs of rain based on observation of the lunar phases, nature of thunder and lightning, direction of sunrise, behaviour of certain plants and animals, and weather forecasts based on the movement of winds; pollenized air and winds; and formation of winds and vapours.
- 1021 - Ibn al-Haytham (Alhazen) introduces the scientific method in his *Book of Optics*. He writes on the atmospheric refraction of light, for example, the cause of morning and evening twilight. He endeavored by use of hyperbola and geometric optics to chart and formulate basic laws on atmospheric refraction. He provides the first correct definition of the twilight, discusses atmospheric refraction, shows that the twilight is due to atmospheric refraction and only begins when the Sun is 19 degrees below the horizon, and uses a complex geometric demonstration to measure the height of the Earth's atmosphere as 52,000 *passuum* (49 miles), which is very close to the modern measurement of 50 miles. He also realized that the atmosphere also reflects light, from his observations of the sky brightening even before the Sun rises.
- 1020s - Ibn al-Haytham publishes his *Risala fi l-Daw'* (*Treatise on Light*) as a supplement to his *Book of Optics*. He discusses the meteorology of the rainbow, the density of the atmosphere, and various celestial phenomena, including the eclipse, twilight and moonlight.
- 1027 - Avicenna publishes *The Book of Healing*, in which Part 2, Section 5, contains his essay on mineralogy and meteorology

in six chapters: formation of mountains; the advantages of mountains in the formation of clouds; sources of water; origin of earthquakes; formation of minerals; and the diversity of earth's terrain. He also describes the structure of a meteor, and his theory on the formation of metals combined Jâbir ibn Hayyân's sulfur–mercury theory from Islamic alchemy (although he was critical of alchemy) with the mineralogical theories of Aristotle and Theophrastus. His scientific methodology of field observation was also original in the Earth sciences.

- Late 11th century - Abu 'Abd Allah Muhammad ibn Ma'udh, who lived in Al-Andalus, wrote a work on optics later translated into Latin as *Liber de crepisculis,* which was mistakenly attributed to Alhazen. This was a short work containing an estimation of the angle of depression of the sun at the beginning of the morning twilight and at the end of the evening twilight, and an attempt to calculate on the basis of this and other data the height of the atmospheric moisture responsible for the refraction of the sun's rays. Through his experiments, he obtained the accurate value of 18°, which comes close to the modern value.
- 1088 - In his *Dream Pool Essays* (¦hªn—{Œ), the Chinese scientist Shen Kuo wrote vivid descriptions of tornadoes, that rainbows were formed by the shadow of the sun in rain, occurring when the sun would shine upon it, and the curious common phenomena of the effect of lightning that, when striking a house, would merely scorch the walls a bit but completely melt to liquid all metal objects inside.
- 1121 - Al-Khazini, a Muslim scientist of Byzantine Greek descent, publishes *The Book of the Balance of Wisdom,* the first study on the hydrostatic balance.
- 13th century-St. Albert the Great is the first to propose that each drop of falling rain had the form of a small sphere, and that this form meant that the rainbow was produced by light interacting with each raindrop.
- Roger Bacon was the first to calculate the angular size of the rainbow. He stated that the rainbow summit can not appear higher than 42 degrees above the horizon.
- Late 13th century-Theoderic of Freiburg and Kamâl al-Dîn al-Fârisî give the first accurate explanations of the primary

rainbow, simultaneously but independently. Theoderic also gives the explanation for the secondary rainbow.

- 1441 - King Sejongs son, Prince Munjong, invented the first standardized rain gauge. These were sent throughout the Joseon Dynasty of Korea as an official tool to assess land taxes based upon a farmer's potential harvest.
- 1450 - Leone Battista Alberti developed a swinging-plate anemometre, and is known as the first *anemometre.*
- Nicolas Cryfts, (Nicolas of Cusa), described the first hair hygrometre to measure humidity. The design was drawn by Leonardo da Vinci, referencing Cryfts design in *da Vinci's Codex Atlanticus.*
- 1488 Johannes Lichtenberger publishes the first version of his *Prognosticatio* linking weather forecasting with astrology. The paradigm was only challenged centuries later.
- 1494 Christopher Columbus experience a tropical cyclone, leads to the first written European account of a hurricane.

2

Fields of Physical Geography

Geomorphology

Geomorphology is the scientific study of landforms and the processes that shape them. Geomorphologists seek to understand why landscapes look the way they do: to understand landform history and dynamics, and predict future changes through a combination of field observation, physical experiment, and numerical modelling. Geomorphology is practiced within geography, geology, geodesy, engineering geology, archaeology, and geotechnical engineering. Early studies in geomorphology are the foundation for pedology, one of two main branches of soil science. More recent studies in geomorphology, pioneered and popularized by Henry Posamentier, integrate seismic geomorphology and seismic stratigraphy, leveraging both 2D and 3D seismic data to better understand the paleogeographic distribution of lithologies.

Landforms evolve in response to a combination of natural and anthropogenic processes. The landscape is built up through tectonic uplift and volcanism. Denudation occurs by erosion and mass wasting, which produces sediment that is transported and deposited elsewhere within the landscape or off the coast. Landscapes are also lowered by subsidence, either due to tectonics or physical changes in underlying sedimentary deposits. These processes are each influenced differently by climate, ecology, and human activity.

Practical applications of geomorphology include hazard assessment including landslide prediction and mitigation, river control and restoration, and coastal protection:

- Hydrology is predominantly concerned with the amounts and quality of water moving and accumulating on the land surface and in the soils and rocks near the surface and is typified by the hydrological cycle. Thus the field encompasses water in rivers, lakes, aquifers and to an extent glaciers, in which the field examines the process and dynamics involved in these bodies of water. Hydrology has historically had an important connection with engineering and has thus developed a largely quantitative method in its research; however, it does have an earth science side that embraces the systems approach. Similar to most fields of physical geography it has sub-fields that examine the specific bodies of water or their interaction with other spheres e.g. limnology and ecohydrology.

- Glaciology is the study of glaciers and ice sheets, or more commonly the cryosphere or ice and phenomena that involve ice. Glaciology groups the latter (ice sheets) as continental glaciers and the former (glaciers) as alpine glaciers. Although, research in the areas are similar with research undertaken into both the dynamics of ice sheets and glaciers the former tends to be concerned with the interaction of ice sheets with the present climate and the latter with the impact of glaciers on the landscape. Glaciology also has a vast array of sub-fields examining the factors and processes involved in ice sheets and glaciers e.g. snow hydrology and glacial geology.
- Biogeography is the science which deals with geographic patterns of species distribution and the processes that result in these patterns. Biogeography emerged as a field of study as a result of the work of Alfred Russel Wallace, although the field prior to the late twentieth century had largely been viewed as historic in its outlook and descriptive in its approach. The main stimulus for the field since its founding has been that of evolution, plate tectonics and the theory of island biogeography. The field can largely be divided into five sub-fields: island biogeography, paleobiogeography, phylogeography, zoogeography and phytogeography
- Climatology is the study of the climate, scientifically defined as weather conditions averaged over a long period of time. It differs from meteorology, which studies atmospheric processes over a shorter duration, which are then examined by climatologists to find trends and frequencies in weather patterns/phenomena. Climatology examines both the nature of micro (local) and macro (global) climates and the natural and anthropogenic influences on them. The field is also sub-divided

largely into the climates of various regions and the study of specific phenomena or time periods e.g. tropical cyclone rainfall climatology and paleoclimatology.

- Pedology is the study of soils in their natural environment. It is one of two main branches of soil science, the other being edaphology. Pedology mainly deals with pedogenesis, soil morphology, soil classification. In physical geography pedology is largely studied due to the numerous interactions between climate (water, air, temperature), soil life (micro-organisms, plants, animals), the mineral materials within soils (biogeochemical cycles) and its position and effects on the landscape such as laterization.
- Palaeogeography is the study of the distribution of the continents through geologic time through examining the preserved material in the stratigraphic record. Palaeogeography is a cross-discipline, almost all the evidence for the positions of the continents comes from geology in the form of fossils or geophysics the use of this data has resulted in evidence for continental drift, plate tectonics and supercontinents this in turn has supported palaeogeographic theories such as the Wilson cycle.
- Coastal geography is the study of the dynamic interface between the ocean and the land, incorporating both the physical geography (i.e coastal geomorphology, geology and oceanography) and the human geography of the coast. It involves an understanding of coastal weathering processes, particularly wave action, sediment movement and weathering, and also the ways in which humans interact with the coast. Coastal geography although predominantly geomorphological in its research is not just concerned with coastal landforms, but also the causes and influences of sea level change.
- Oceanography is the branch of physical geography that studies the Earth's oceans and seas. It covers a wide range of topics, including marine organisms and ecosystem dynamics (biological oceanography); ocean currents, waves, and geophysical fluid dynamics (physical oceanography); plate tectonics and the geology of the sea floor (geological oceanography); and fluxes of various chemical substances and physical properties within the ocean and across its boundaries (chemical oceanography). These diverse topics reflect multiple disciplines that

oceanographers blend to further knowledge of the world ocean and understanding of processes within it.

- Quaternary science is an inter-disciplinary field of study focusing on the Quaternary period, which encompasses the last 2.6 million years. The field studies the last ice age and the recent interstadial the Holocene and uses proxy evidence to reconstruct the past environments during this period to infer the climatic and environmental changes that have occurred.
- Landscape ecology is a sub-discipline of ecology and geography that address how spatial variation in the landscape affects ecological processes such as the distribution and flow of energy, materials and individuals in the environment (which, in turn, may influence the distribution of landscape "elements" themselves such as hedgerows). The field was largely founded by the German geographer Carl Troll Landscape ecology typically deals with problems in an applied and holistic context. The main difference between biogeography and landscape ecology is that the latter is concerned with how flows or energy and material are changed and their impacts on the landscape whereas the former is concerned with the spatial patterns of species and chemical cycles.
- Geomatics is the field of gathering, storing, processing, and delivering of geographic information, or spatially referenced information. Geomatics includes geodesy (scientific discipline that deals with the measurement and representation of the earth, its gravitational field, and other geodynamic phenomena, such as crustal motion, oceanic tides, and polar motion) and G.I.S. (a system for capturing, storing, analyzing and managing data and associated attributes which are spatially referenced to the earth) and remote sensing (the short or large-scale acquisition of information of an object or phenomenon, by the use of either recording or real-time sensing devices that are not in physical or intimate contact with the object).
- Environmental geography is a branch of geography that analyzes the spatial aspects of interactions between humans and the natural world. The branch bridges the divide between human and physical geography and thus requires an understanding of the dynamics of geology, meteorology, hydrology, biogeography, and geomorphology, as well as the

ways in which human societies conceptualize the environment. Although the branch was previously more visible in research than at present with theories such as environmental determinism linking society with the environment. It has largely become the domain of the study of environmental management or anthropogenic influences on the environment and vice a versa.

Physical Geography of India

The geography of India describes the physical features of India, a country in South Asia that lies entirely on the Indian Plate in the northern portion of the Indo-Australian Plate. The country lies to the north of the equator between 8°4' and 37°6' north latitude and 68°7' and 97°25' east longitude. It is the seventh-largest country in the world, with a total land area of 3,287,263 square kilometres (1,269,219 sq mi). India measures 3,214 km (1,997 mi) from north to south and 2,993 km (1,860 mi) from east to west. It has a land frontier of 15,200 km (9,445 mi) and a coastline of 7,517 km (4,671 mi).

India is bounded to the southwest by the Arabian Sea, to the southeast by the Bay of Bengal and the Indian Ocean to the south. Kanyakumari constitutes the southern tip of the Indian peninsula, which narrows before ending in the Indian Ocean. The southernmost part of India is Indira Point in the Andaman and Nicobar Islands. The Maldives, Sri Lanka and Indonesia are island nations to the south of India with Sri Lanka separated from India by a narrow channel of sea formed by Palk Strait and the Gulf of Mannar. The territorial waters of India extend into the sea to a distance of 12 nautical miles (13.8 mi; 22.2 km) measured from the appropriate baseline.

The northern frontiers of India are defined largely by the Himalayan mountain range where its political boundaries with China, Bhutan, and Nepal lie. Its western borders with Pakistan lie in the Punjab Plain and the Thar desert. In the far northeast, the Chin Hills and Kachin Hills, deeply forested mountainous regions, separate India from Burma while its political border with Bangladesh is defined by the watershed region of the Indo-Gangetic Plain, the Khasi hills and Mizo Hills.

The Ganges is the longest river originating in India and forms the Indo-Gangetic Plain. The Ganges-Brahmaputra system occupies most of northern, central and eastern India, while the Deccan Plateau occupies most of southern India. Along its western frontier is the Thar Desert, which is the seventh-largest desert in the world.

Officially, India's highest point is K2 at 8,611 m (28,251 ft), though it lies in Gilgit-Baltistan, part of the disputed Kashmir region. Kanchenjunga in Sikkim at 8,598 m (28,209 ft) is the highest point within India's current geographic boundaries. Climate across India ranges from equatorial in the far south, to Alpine in the upper reaches of the Himalayas.

Geological Development

India is entirely contained on the Indian Plate, a major tectonic plate that was formed when it split off from the ancient continent Gondwanaland. About 90 million years ago, during the late Cretaceous Period, the Indian Plate began moving north at about 15 cm/yr (6 in/yr). About 50 to 55 million years ago, in the Eocene epoch of the Cenozoic Era, the plate collided with Asia after covering a distance of 2,000 to 3,000 km (1,243 to 1,864 mi), having moved faster than any other known plate. In 2007, German geologists determined that the reason the India Plate moved so quickly is that it is only half as thick as the other plates which formerly constituted Gondwanaland. The collision with the Eurasian Plate along the modern border between India and Nepal formed the orogenic belt that created the Tibetan Plateau and the Himalayas. As of 2009, The India Plate is moving northeast at 5 cm/yr (2 in/yr), while the Eurasian Plate is moving north at only 2 cm/yr (0.8 in/yr). India is thus referred to as the "fastest continent." This is causing the Eurasian Plate to deform, and the India Plate to compress at a rate of 4 mm/yr (0.15 in/yr).

Political Geography

India is divided into twenty-eight states (further subdivided into districts) and seven union territories.

States:

1. Andhra Pradesh
2. Arunachal Pradesh
3. Assam
4. Bihar
5. Chhattisgarh
6. Goa
7. Gujarat
8. Haryana
9. Himachal Pradesh
10. Jammu and Kashmir
11. Jharkhand
12. Karnataka
13. Kerala
14. Madhya Pradesh
15. Maharashtra
16. Manipur
17. Meghalaya
18. Mizoram
19. Nagaland
20. Orissa

21. Punjab
22. Rajasthan
23. Sikkim
24. Tamil Nadu
25. Tripura
26. Uttar Pradesh
27. Uttarakhand
28. West Bengal

Union Territories:

A. Andaman and Nicobar Islands
B. Chandigarh
C. Dadra and Nagar Haveli
D. Daman and Diu
E. Lakshadweep
F. National Capital Territory of Delhi
G. Puducherry.

India's borders run a total length of 15,106.70 km (9,387 mi). Its borders with Pakistan and Bangladesh were delineated according to the Radcliffe Line, which was created in 1947 during Partition of India. Its western border with Pakistan extends up to 3,323 km (2,065 mi), dividing the Punjab region and running along the boundaries of the Thar Desert and the Rann of Kutch.

Both nations delineated a Line of Control (LoC) to serve as the informal boundary between the Indian and Pakistan-administered areas of Kashmir. According to India's claim, it shares a 106 km (66 mi) border with Afghanistan in northwestern Kashmir, which is under Pakistani control.

India's border with Bangladesh runs 4,096.70 km (2,546 mi). There are 92 enclaves of Bangladesh on Indian soil and 106 enclaves of India are on Bangladeshi soil. The Teen Bigha Corridor is a strip of land formerly belonging to India on the West Bengal–Bangladesh border which has been leased indefinitely to Bangladesh so that it can access its Dehgram–Angalpota enclaves.

The Line of Actual Control (LAC) is the effective border between India and the People's Republic of China. It traverses 4,057 km along the Indian states of Jammu and Kashmir, Uttarakhand, Himachal Pradesh, Sikkim and Arunachal Pradesh. Both nations lay claim to the Aksai Chin region of northeastern Kashmir, which fell into Chinese control during the Sino-Indian War of 1962.

The border with Burma (Myanmar) extends up to 1,643 km (1,021 mi) along the southern borders of India's northeastern states. Located amidst the Himalayan range, India's border with Bhutan runs 699 km (434 mi). The border with Nepal runs 1,751 km (1,088 mi) along the foothills of the Himalayas in northern India. The Siliguri Corridor, narrowed sharply by the borders of Bhutan, Nepal and Bangladesh, connects peninsular India with the northeastern states.

Physiographic Regions

India is divided into seven physiographic regions. They are

1. The northern mountains including the Himalayas, which includes the Kuen Lun and the Karakoram ranges and the northeast mountain ranges.
2. Indo-Gangetic plains
3. Thar Desert
4. Central Highlands and Deccan Plateau
5. East Coast
6. West Coast
7. Bordering seas and islands.

Mountains

A great arc of mountains, consisting of the Himalayas, Hindu Kush, and Patkai ranges define the northern Indian subcontinent. These were formed by the ongoing tectonic collision of the Indian Plate with the Eurasian Plate that started around 50 million years ago. The mountains in these ranges include some of the world's tallest mountains which act as a natural barrier to cold polar winds. They also facilitate the monsoon winds which in turn influence the climate in India. Rivers originating in these mountains, flow through the fertile Indo–Gangetic plains. These mountains are recognised by biogeographers as the boundary between two of the Earth's great ecozones: the temperate Palearctic that covers most of Eurasia and the tropical and subtropical Indomalaya ecozone which includes the Indian subcontinent, Southeast Asia and Indonesia.

India has eight major mountain ranges having peaks of over 1,000 m (3,281 ft):

- The Himalayan range is considered as the world's highest mountain range, with its tallest peak Mt. Everest on the Nepal–China border. They form India's northeastern border, separating it from northeastern Asia. They are one of the world's youngest mountain ranges and extend almost uninterrupted for 2,500 km (1,553 mi), covering an area of 500,000 km^2 (193,051 sq mi). The Himalayas extend from

Jammu and Kashmir in the west to Arunachal Pradesh in the east. These states along with Himachal Pradesh, Uttarakhand, and Sikkim lie mostly in the Himalayan region. Numerous Himalayan peaks rise over 7,000 m (22,966 ft) and the snow line ranges between 6,000 m (19,685 ft) in Sikkim to around 3,000 m (9,843 ft) in Kashmir. Kanchenjunga—on the Sikkim–Nepal border—is the highest point in the area administered by India. Most peaks in the Himalayas remain snowbound throughout the year. The Himalayas act as a barrier to the frigid katabatic winds flowing down from Central Asia. Thus, North India is kept warm or only mildly cooled during winter; in summer, the same phenomenon makes India relatively hot.

- The Karakoram is situated in the disputed state of Jammu and Kashmir. It has more than sixty peaks above 7,000 m (22,966 ft), including K2, the second highest peak in the world 8,611 m (28,251 ft). K2 is just 237 m (778 ft) smaller than the 8,848 m (29,029 ft) Mount Everest. The range is about 500 km (311 mi) in length and the most heavily glaciated part of the world outside of the polar regions. The Siachen Glacier at 70 km (43 mi) and the Biafo Glacier at 63 km (39 mi) rank as the world's second and third-longest glaciers outside the polar regions. Just to the west of the northwest end of the Karakoram, lies the Hindu Raj range, beyond which is the Hindu Kush range. The southern boundary of the Karakoram is formed by the Gilgit, Indus and Shyok rivers, which separate the range from the northwestern end of the Himalayas.
- The Patkai, or Purvanchal, are situated near India's eastern border with Myanmar. They were created by the same tectonic processes which led to the formation of the Himalayas. The physical features of the Patkai mountains are conical peaks, steep slopes and deep valleys. The Patkai ranges are not as rugged or tall as the Himalayas. There are three hill ranges that come under the Patkai: the Patkai–Bum, the Garo–Khasi–Jaintia and the Lushai hills. The Garo–Khasi range lies in Meghalaya. Mawsynram, a village near Cherrapunji lying on the windward side of these hills, has the distinction of being the wettest place in the world, receiving the highest annual rainfall.
- The Vindhya range runs across most of central India, extending 1,050 km (652 mi). The average elevation of these hills is 3,000 m (9,843 ft). They are believed to have been formed by

the wastes created by the weathering of the ancient Aravali mountains. Geographically, it separates northern India from southern India. The western end of the range lies in eastern Gujarat, near its border with Madhya Pradesh, and runs east and north, almost meeting the Ganges at Mirzapur.

- The Satpura Range begins in eastern Gujarat near the Arabian Sea coast and runs east across Maharashtra, Madhya Pradesh and Chhattisgarh. It extends 900 km (559 mi) with many peaks rising above 1,000 m (3,281 ft). It is triangular in shape, with its apex at Ratnapuri and the two sides being parallel to the Tapti and Narmada rivers. It runs parallel to the Vindhya Range, which lies to the north, and these two east-west ranges divide the Indo–Gangetic plain from the Deccan Plateau located north of River Narmada.
- The Aravali Range is the oldest mountain range in India, running across Rajasthan from northeast to southwest direction, extending approximately 800 km (497 mi). The northern end of the range continues as isolated hills and rocky ridges into Haryana, ending near Delhi. The highest peak in this range is Guru Shikhar at Mount Abu, rising to 1,722 m (5,650 ft), lying near the border with Gujarat. The Aravali Range is the eroded stub of an ancient fold mountain system. The range rose in a Precambrian event called the Aravali–Delhi orogen. The range joins two of the ancient segments that make up the Indian craton, the Marwar segment to the northwest of the range, and the Bundelkhand segment to the southeast.
- The Western Ghats or Sahyadri mountains run along the western edge of India's Deccan Plateau and separate it from a narrow coastal plain along the Arabian Sea. The range runs approximately 1,600 km (994 mi) from south of the Tapti River near the Gujarat–Maharashtra border and across Maharashtra, Goa, Karnataka, Kerala and Tamil Nadu to the southern tip of the Deccan peninsula. The average elevation is around 1,000 m (3,281 ft). Anai Mudi in the Anaimalai Hills 2,695 m (8,842 ft) in Kerala is the highest peak in the Western Ghats.
- The Eastern Ghats are a discontinuous range of mountains, which have been eroded and vivisected by the four major rivers of southern India, the Godavari, Mahanadi, Krishna, and Kaveri. These mountains extend from West Bengal to Orissa, Andhra Pradesh and Tamil Nadu, along the coast and parallel to the Bay of Bengal. Though not as tall as the Western Ghats,

some of its peaks are over 1,000 m (3,281 ft) in height. The Nilgiri hills in Tamil Nadu lies at the junction of the Eastern and Western Ghats.

Indo-Gangetic Plain

The Indo-Gangetic plains, also known as the *Great Plains* are large floodplains of the Indus and the Ganga-Brahmaputra river systems. They run parallel to the Himalaya mountains, from Jammu and Kashmir in the west to Assam in the east and draining most of northern and eastern India. The plains encompass an area of 700,000 square kilometres (270,272 sq mi). The major rivers in this region are the Ganges and the Indus along with their tributaries–Beas, Yamuna, Gomti, Ravi, Chambal, Sutlej and Chenab.

The great plains are sometimes classified into four divisions:

- The Bhabar belt — is adjacent to the foothills of the Himalayas and consists of boulders and pebbles which have been carried down by the river streams. As the porosity of this belt is very high, the streams flow underground. The bhabar is generally narrow with its width varying between 7 to 15 km.
- The Terai belt — lies next to the Bhabar region and is composed of newer alluvium. The underground streams reappear in this region. The region is excessively moist and thickly forested. It also receives heavy rainfall throughout the year and is populated with a variety of wildlife.
- The Bangar belt — consists of older alluvium and forms the alluvial terrace of the flood plains. In the Gangetic plains, it has a low upland covered by laterite deposits.
- The Khadar belt — lies in lowland areas after the Bangar belt. It is made up of fresh newer alluvium which is deposited by the rivers flowing down the plain.

The Indo-Gangetic belt is the world's most extensive expanse of uninterrupted alluvium formed by the deposition of silt by the numerous rivers. The plains are flat making it conducive for irrigation through canals. The area is also rich in ground water sources.

The plains are one of the world's most intensely farmed areas. The main crops grown are rice and wheat, which are grown in rotation. Other important crops grown in the region include maize, sugarcane and cotton. The Indo-Gangetic plains rank among the world's most densely populated areas.

Thar Desert

The Thar Desert (also known as the *Great Indian Desert*) is the world's seventh largest desert; it forms a significant portion of western India and covers an area of 238,700 km^2 (92,200 $mile^2$). The desert continues into Pakistan as the Cholistan Desert. Most of the Thar Desert is situated in Rajasthan, covering 61% of its geographic area.

About 10 percent of this ecoregion comprises sand dunes, and the remaining 90 percent consist of craggy rock forms, compacted salt-lake bottoms, and interdunal and fixed dune areas. Annual temperatures can range from 0°C in the winter to over 50°C during the summer. Most of the rainfall received in this region is associated with the short July-September southwest monsoon that brings around 100–500 mm of precipitation. Water is scarce and occurs at great depths, ranging from 30 to 120 m below the ground level. Rainfall is precarious and erratic, ranging from below 120 mm (4.72 inches) in the extreme west to 375 mm (14.75 inches) eastward. The soils of the arid region are generally sandy to sandy-loam in texture. The consistency and depth vary as per the topographical features. The low-lying loams are heavier and may have a hard pan of clay, calcium carbonate or gypsum.

Highlands

The Central Highlands comprise of three main plateaus — the Malwa Plateau in the west, the Deccan Plateau in the south (covering most of the Indian peninsula) and the Chota Nagpur Plateau in the east.

The Malwa Plateau is spread across Rajasthan, Madhya Pradesh and Gujarat. The average elevation of the Malwa plateau is 500 metres, and the landscape generally slopes towards the north. Most of the region is drained by the Chambal River and its tributaries; the western part is drained by the upper reaches of the Mahi River. The Deccan Plateau is a large triangular plateau, bounded by the Vindhyas to the north and flanked by the Eastern and Western Ghats. The Deccan covers a total area of 1.9 million km^2 (735,000 $mile^2$). It is mostly flat, with elevations ranging from 300 to 600 m (1,000 to 2,000 ft). The average elevation of the plateau is 2,000 feet (600 m) above sea level. The surface slopes from 3,000 feet (900 m) in the west to 1,500 feet (450 m) in the east. It slopes gently from west to east and gives rise to several peninsular rivers such as the Godavari, the Krishna, the Kaveri and the Narmada, which drain into the Bay of Bengal. This region is mostly semi-arid as it lies on the leeward side of both Ghats. Much of the Deccan is covered by thorn scrub forest scattered with small regions of deciduous broadleaf forest. Climate in the Deccan ranges from hot summers to mild winters.

The Chota Nagpur Plateau is situated in eastern India, covering much of Jharkhand and adjacent parts of Orissa, Bihar and Chhattisgarh. Its total area is approximately 65,000 km^2 (25,000 mile^2) and is made up of three smaller plateaus — the Ranchi, Hazaribagh, and Kodarma plateaus. The Ranchi plateau is the largest, with an average elevation of 700 m (2,300 ft). Much of the plateau is forested, covered by the Chota Nagpur dry deciduous forests. Vast reserves of metal ores and coal have been found in the Chota Nagpur plateau. The Kathiawar peninsula in western Gujarat is bounded by the Gulf of Kutch and the Gulf of Khambat. The natural vegetation in most of the peninsula is xeric scrub, part of the Northwestern thorn scrub forests ecoregion. In western India, the Kutch region in Gujarat and Koyna in Maharashtra are classified as a Zone IV region (high risk) for earthquakes. The Kutch city of Bhuj was the epicentre of the 2001 Gujarat earthquake, which claimed the lives of more than 20,000 people and injured 166,836 while destroying or damaging near a million homes. The 1993 Latur earthquake in Maharashtra killed 7,928 people and injured 30,000. Other areas have a moderate to low risk of an earthquake occurring.

Coasts

The Eastern Coastal Plain is a wide stretch of land lying between the Eastern Ghats and the Bay of Bengal. It stretches from Tamil Nadu in the south to West Bengal in the north. The Mahanadi, Godavari, Kaveri and Krishna rivers drain these plains and their deltas occupy most of the area. The temperature in the coastal regions exceeds 30 °C (86 °F) coupled with high levels of humidity. The region receives both the northeast and southwest monsoon rains. The southwest monsoon splits into two branches, the Bay of Bengal branch and the Arabian Sea branch. The Bay of Bengal branch moves northwards crossing northeast India in early June. The Arabian Sea branch moves northwards and discharges much of its rain on the windward side of Western Ghats. Annual rainfall in this region averages between 1,000 mm (40 in) and 3,000 mm (120 in). The width of the plains varies between 100 and 130 km (62 to 80 miles). The plains are divided into six regions — the Mahanadi delta, the southern Andhra Pradesh plain, the Krishna-Godavari deltas, the Kanyakumari coast, the Coromandel Coast and sandy coastal.

The Western Coastal Plain is a narrow strip of land sandwiched between the Western Ghats and the Arabian Sea, ranging from 50 to 100 km (30 to 60 miles) in width. It extends from Gujarat in the north and extends through Maharashtra, Goa, Karnataka and Kerala. Numerous rivers and backwaters inundate the region. Originating in the Western Ghats, the rivers are fast-flowing and mostly perennial, leading to the formation of estuaries. Major rivers flowing into the sea are the Tapi, Narmada, Mandovi and Zuari. The coast is divided into 3 parts namely, Konkan, which is situated in Maharashtra, Goa and northern parts of Karnataka; the Kanara in Karnataka and the Malabar Coast in Kerala. Vegetation is mostly deciduous, but the Malabar Coast moist forests constitute a unique ecoregion.

Islands

The Lakshadweep and the Andaman and Nicobar Islands are India's two major island formations which are classified as union territories. The Lakshadweep Islands lie 200 to 300 km (124 to 186 miles) off the coast of Kerala in the Arabian Sea with an area of 32 km^2 (11 sq mi). They consist of 12 atolls, 3 reefs and 5 submerged banks, with a total of about 36 islands and islets.

The Andaman and Nicobar Islands are located between 6° and 14° North latitude and 92° and 94° East longitude. They consist of 572 isles, lying in the Bay of Bengal near the Myanmar coast. It is located 1255 km (780 miles) from Kolkata (Calcutta) and 193 km (120 miles) from Cape Negrais in Myanmar. The territory consists of two island groups, the Andaman Islands and the Nicobar Islands. The Andaman Islands consists of 204 small islands with a total length of 352 km (220 miles). India's only active volcano, Barren Island is situated here, having last erupted in May 2005. The Narcondum is a dormant volcano and there is a mud volcano at Baratang. Indira Point, India's southernmost land point is situated in the Nicobar islands, and lies just 189 km (117 miles) from the Indonesian island of Sumatra to the southeast. The highest point is Mount Thullier at 642 m (2,140 ft).

Significant islands just off the Indian coast include Diu, a former Portuguese enclave; Majuli, Asia's largest freshwater island; Elephanta in the Bombay Harbour; and Sriharikota barrier island in Andhra Pradesh. Salsette Island is India's most populous island on which the city of Mumbai (Bombay) is located. Forty-two islands in the Gulf of Kutch constitute the Marine National Park.

Water Bodies

India has around 14,500 km of inland navigable waterways. There are twelve rivers which are classified as major rivers, with the total catchment area exceeding 2,528,000 km^2 (976,000 sq mi). All major rivers of India originate from one of the three main watersheds:

1. The Himalaya and the Karakoram ranges
2. Vindhya and Satpura range in central India
3. Sahyadri or Western Ghats in western India.

The Himalayan river networks are snow-fed and have a perennial supply throughout the year. The other two river systems are dependent on the monsoons and shrink into rivulets during the dry season. The Himalayan rivers that flow westward into Pakistan are the Indus, Beas, Chenab, Ravi, Sutlej, and Jhelum.

The Ganga-Brahmaputra-Meghana system has the largest catchment area of 1,100,000 km^2 (420,000 sq mi). The Ganga originates from the Gangotri Glacier in Uttarakhand. It flows southeast, draining into the Bay of Bengal. The Yamuna and Gomti rivers also arise in the western Himalayas and join the Ganga in the plains. The Brahmaputra, another tributary of the Ganga, originates in Tibet and enters India through the far-eastern state of Arunachal Pradesh. It proceeds westwards, joining the Ganges in Bangladesh.

The Chambal, another tributary of the Ganga originates from the Vindhya-Satpura watershed. The river flows eastward. Westward-flowing rivers from this watershed are the Narmada and Tapti, which drain into the Arabian Sea in Gujarat. The river network that flows from east to west constitutes 10% of the total outflow.

The Western Ghats are the source of all Deccan rivers, which include the Mahanadi River through the Mahanadi River Delta, Godavari River, Krishna River and Kaveri River, all draining into the Bay of Bengal. These rivers constitute 20% of India's total outflow.

The heavy southwest monsoon rains cause the Brahmaputra and other rivers to distend their banks, often flooding surrounding areas. Though they provide rice paddy farmers with a largely dependable source of natural irrigation and fertilisation, such floods have killed thousands of people and tend to cause displacements of people in such areas. Major gulfs include the Gulf of Cambay, Gulf of Kutch and the Gulf of Mannar. Straits include the Palk Strait, which separates India from Sri Lanka and the Ten Degree Channel, which separates the Andamans from the Nicobar Islands and the Eight Degree Channel, which separates the Laccadive and Amindivi Islands from the Minicoy Island towards the south. Important capes include the Kanyakumari, the southern tip of mainland India; Indira Point, the southernmost location of India; Rama's Bridge and Point Calimere. While, Arabian Sea lies on the western side of India, Bay of Bengal and Indian Ocean lie towards the eastern and southern side respectively. Smaller seas include the Laccadive Sea and the Andaman Sea. There are four coral reefs in India, located in the Andaman and Nicobar Islands, Gulf of Mannar, Lakshadweep and Gulf of Kutch. Important lakes include Chilka Lake, the country's largest saltwater lake in Orissa; Kolleru Lake in Andhra Pradesh; Loktak Lake in Manipur, Dal Lake in Kashmir, Sambhar Lake in Rajasthan and the Sasthamkotta Lake in Kerala.

Wetlands

India's wetland ecosystem is widely distributed from the cold and arid located in the Ladakh region of Jammu and Kashmir, and those with the wet and humid climate of peninsular India. Most of the wetlands are directly or indirectly linked to river networks. The Indian government has identified a total of 71 wetlands for conservation and are part of sanctuaries and national parks. Mangrove forests are present all along the Indian coastline in sheltered estuaries, creeks, backwaters, salt marshes and mudflats. The mangrove area covers a total of 4,461 km^2 (1,722 sq mi), which comprises 7% of the world's total mangrove cover. Prominent mangrove covers are located in the Andaman and Nicobar Islands, the Sundarbans delta, the Gulf of Kutch and the deltas of the Mahanadi, Godavari and Krishna rivers. Parts of Maharashtra, Karnataka and Kerala also have large mangrove covers.

The Sundarbans delta is home to the largest mangrove forest in the world. It lies at the mouth of the Ganges and spreads across areas of Bangladesh and West Bengal. The Sundarbans is a UNESCO World Heritage Site, but is identified separately as the Sundarbans (Bangladesh) and the Sundarbans National Park (India). The Sundarbans are intersected by a complex network of tidal waterways, mudflats and small islands of salt-tolerant mangrove forests. The area is known for its diverse fauna, being home to a large variety of species of birds, spotted deer, crocodiles and snakes. Its most famous inhabitant is the Bengal Tiger. It is estimated that there are now 400 Bengal tigers and about 30,000 spotted deer in the area.

The Rann of Kutch is a marshy region located in northwestern Gujarat and the bordering Sindh province of Pakistan. It occupies a total area of 27 900 km^2 (10,800 mile^2). The region was originally a part of the Arabian Sea. Geologic forces such as earthquakes resulted in the damming up of the region, turning it into a large saltwater lagoon. This area gradually filled with silt thus turning it into a seasonal salt marsh. During the monsoons, the area turns into a shallow marsh, often flooding to knee-depth. After the monsoons, the region turns dry and becomes parched.

Climate

Based on the Koppen system, India hosts six major climatic subtypes, ranging from arid desert in the west, alpine tundra and glaciers in the north, and humid tropical regions supporting rainforests in the southwest and the island territories. Many regions have starkly different microclimates. The nation has four seasons: winter (January–February), summer (March–May), a monsoon (rainy) season (June–September) and a post-monsoon period (October–December).

The Himalayas act as a barrier to the frigid katabatic winds flowing down from Central Asia. Thus, North India is kept warm or only mildly cooled during winter; in summer, the same phenomenon makes India relatively hot. Although the Tropic of Cancer—the boundary between the tropics and subtropics—passes through the middle of India, the whole country is considered to be tropical.

Summer lasts between March and June in most parts of India. Temperatures exceed 40 °C (104 °F) during the day. The coastal regions exceed 30 °C (86 °F) coupled with high levels of humidity. In the Thar desert area temperatures can exceed 45 °C (113 °F). The rain-bearing monsoon clouds are attracted to the low-pressure system created by the Thar Desert. The southwest monsoon splits into two arms, the Bay of Bengal arm and the Arabian Sea arm. The Bay of Bengal arm moves northwards crossing northeast India in early June. The Arabian Sea arm moves northwards and deposits much of its rain on the windward side of Western Ghats. Winters in peninsula India see mild to warm days and cool nights. Further north the temperature is cooler. Temperatures in some parts of the Indian plains sometimes fall below freezing. Most of northern India is plagued by fog during this season. The highest temperature recorded in India was 50.6 °C (123.1 °F) in Alwar in 1955. The lowest was "45 °C ("49.0 °F) in Kashmir.

Geology

India's geological features are classified based on their era of formation. The Precambrian formations of Cudappah and Vindhyan systems are spread out over the eastern and southern states. A small part of this period is spread over western and central India. The Paleozoic formations from the Cambrian, Ordovician, Silurian and Devonian system are found in the Western Himalaya region in Kashmir and Himachal Pradesh. The Mesozoic Deccan Traps formation is seen over most of the northern Deccan; they are believed to be the result of sub-aerial volcanic activity. The Trap soil is black in colour and conducive to agriculture. The Carboniferous system, Permian System and Triassic systems are seen in the western Himalayas. The Jurassic system is seen in the western Himalayas and Rajasthan. Tertiary imprints are seen in parts of Manipur, Nagaland, Arunachal Pradesh and along the Himalayan belt. The Cretaceous system is seen in central India in the Vindhyas and part of the Indo-Gangetic plains. The Gondowana system is seen in the Narmada River area in the Vindhyas and Satpuras. The Eocene system is seen in the western Himalayas and Assam. Oligocene formations are seen in Kutch and Assam. The Pleistocene system is found over central India. The Andaman and Nicobar Island are thought to have been formed in this era by volcanoes. The Himalayas were formed by the convergence and deformation of the Indo-Australian and Eurasian Plates. Their continued convergence raises the height of the Himalayas by 1 cm each year.

Soils in India can be classified into 8 categories: alluvial, black, red, laterite, forest, arid & desert, saline & alkaline and peaty & organic soils. Alluvial soil constitute the largest soil group in India, constituting 80% of the total land surface. It is derived from the deposition of silt carried by rivers and are found in the Great Northern plains from Punjab to the Assam valley. Alluvial soil are generally fertile but they lack nitrogen and tend to be phosphoric.

Black soil are well developed in the Deccan lava region of Maharashtra, Gujarat, and Madhya Pradesh. These contain high percentage of clay and are moisture retentive. Red soil are found in Tamil Nadu, Karnataka plateau, Andhra plateau, Chota Nagpur plateau and the Aravallis. These are deficient in nitrogen, phosphorus and humus. Laterite soils are formed in tropical regions with heavy rainfall. Heavy rainfall results in leaching out all soluble material of top layer of soil. These are generally found in Western ghats, Eastern ghats and hilly areas of northeastern states that receive heavy rainfall. Forest soils occur on the slopes of mountains and hills in Himalayas, Western Ghats and Eastern Ghats. These generally consist of large amounts of dead leaves and other organic matter called humus.

Natural Resources

India's total renewable water resources are estimated at 1,907.8 km^3/year. Its annual supply of usable and replenshable groundwater amounts to 350 billion cubic metres. Only 35% of groundwater resources are being utilised. About 44 million tonnes of cargo is moved annually through the country's major rivers and waterways. Groundwater supplies 40% of water in India's irrigation canals. 56% of the land is arable and used for agriculture. Black soils are moisture-retentive and are preferred for dry farming and growing cotton, linseed, etc. Forest soils are used for tea and coffee plantations. Red soil have a wide diffusion of iron content.

Most of India's estimated 5.4 billion barrels (860,000,000 m^3) in oil reserves are located in the Mumbai High, upper Assam, Cambay, the Krishna-Godavari and Cauvery basins. India possesses about seventeen trillion cubic feet of natural gas in Andhra Pradesh, Gujarat and Orissa. Uranium is mined in Andhra Pradesh. India has 400 medium-to-high enthalpy thermal springs for producing geothermal energy in seven "provinces" — the Himalayas, Sohana, Cambay, the Narmada-Tapti delta, the Godavari delta and the Andaman and Nicobar Islands (specifically the volcanic Barren Island.)

India is the world's biggest producer of mica blocks and mica splittings. India ranks second amongst the world's largest producers of barites and chromites. The Pleistocene system is rich in minerals. India is the third-largest coal producer in the world and ranks fourth in the production of iron ore. It is the fifth-largest producer of bauxite and crude steel, the seventh-largest of manganese ore and the eighth-largest of aluminium. India has significant sources of titanium ore, diamonds and limestone. India possesses 24% of the world's known and economically-viable thorium, which is mined along shores of Kerala. Gold had been mined in the now-defunct Kolar Gold Fields in Karnataka.

3

The Study of Ecosystems

Ecosystem Dynamics

Introduction of new elements, whether biotic or abiotic, into an ecosystem tend to have a disruptive effect. In some cases, this can lead to ecological collapse or "trophic cascading" and the death of many species within the ecosystem.

Under this deterministic vision, the abstract notion of ecological health attempts to measure the robustness and recovery capacity for an ecosystem; i.e. how far the ecosystem is away from its steady state.

Often, however, ecosystems have the ability to rebound from a disruptive agent. The difference between collapse or a gentle rebound is determined by two factors—the toxicity of the introduced element and the resiliency of the original ecosystem.

Ecosystems are primarily governed by stochastic (chance) events, the reactions these events provoke on non-living materials and the responses by organisms to the conditions surrounding them.

Thus, an ecosystem results from the sum of individual responses of organisms to stimuli from elements in the environment. The presence or absence of populations merely depends on reproductive and dispersal success, and population levels fluctuate in response to stochastic events.

As the number of species in an ecosystem is higher, the number of stimuli is also higher. Since the beginning of life organisms have survived continuous change through natural selection of successful feeding, reproductive and dispersal behaviour. Through natural selection the planet's species have continuously adapted to change through variation in their biological composition and distribution. Mathematically it can be demonstrated that greater numbers of different interacting factors tend to dampen fluctuations in each of the individual factors.

Given the great diversity among organisms on earth, most ecosystems only changed very gradually, as some species would disappear while others would move in. Locally, sub-populations continuously go extinct, to be replaced later through dispersal of other sub-populations. Stochastists do recognise that certain intrinsic regulating mechanisms occur in nature. Feedback and response mechanisms at the species level regulate population levels, most notably through territorial behaviour.

Andrewatha and Birch suggest that territorial behaviour tends to keep populations at levels where food supply is not a limiting factor. Hence, stochastists see territorial behaviour as a regulatory mechanism at the species level but not at the ecosystem level. Thus, in their vision, ecosystems are not regulated by feedback and response mechanisms from the ecosystem itself and there is no such thing as a balance of nature.

If ecosystems are governed primarily by stochastic processes, through which its subsequent state would be determined by both predictable and random actions, they may be more resilient to sudden change than each species individually. In the absence of a balance of nature, the species composition of ecosystems would undergo shifts that would depend on the nature of the change, but entire ecological collapse would probably be infrequent events.

The theoretical ecologist Robert Ulanowicz has used information theory tools to describe the structure of ecosystems, emphasizing mutual information (correlations) in studied systems. Drawing on this methodology and prior observations of complex ecosystems, Ulanowicz depicts approaches to determining the stress levels on ecosystems and predicting system reactions to defined types of alteration in their settings (such as increased or reduced energy flow, and eutrophication.

In addition, Eric Sanderson has developed the Muir web, based on experience on the Mannahatta project. This graphical schematic shows how different species are connected to each other, not only regarding their position in the food chain, but also regarding other services, i.e. provisioning of shelter, ...

Ecosystem Ecology

Ecosystem ecology is the integrated study of biotic and abiotic components of ecosystems and their interactions within an ecosystem framework. This science examines how ecosystems work and relates this to their components such as chemicals, bedrock, soil, plants, and animals. Ecosystem ecology examines physical and biological structure and examines how these ecosystem characteristics interact.

Bio-Diversity and its Depletion Through Natural and Man Induced Causes

Habitat destruction is the process in which natural habitat is rendered functionally unable to support the species present. In this process, the organisms which previously used the site are displaced or destroyed, reducing biodiversity. Habitat destruction by human activity mainly for the purpose of harvesting natural resources for industry production and urbanization. Clearing habitats for agriculture is the principal cause of habitat destruction. Other important causes of habitat destruction include mining, logging, trawling and urban sprawl. Habitat destruction is currently ranked as the primary cause of species extinction worldwide. It is a process of natural environmental change that may be caused by habitat fragmentation, geological processes, climate change or by human activities such as the in troduction of invasive species, ecosystem nutrient d epletion and other human activities mentioned below.

The terms "habitat loss" and "habitat reduction" are also used in a wider sense including loss of habitat from other factors, such as water and noise pollution.

Impacts on Organisms

In the simplest terms, when a habitat is destroyed, the plants, animals, and other organisms that occupied the habitat have a reduced carrying capacity so that populations decline and extinction becomes more likely. Perhaps the greatest threat to organisms and biodiversity is the process of habitat loss. Temple (1986) found that 82% of endangered bird species were significantly threatened by habitat loss. Endemic organisms that obtains limited ranges are most affected by habitat destruction, mainly because these organisms are not found anywhere else within the world and thus, have less chance of recovering. Many endemic organisms have very specific requirements for their survival that can only be found within a certain ecosystem, resulting in their extinction. Extinction may take place very long after the destruction of habitat, though, a phenomenon known as extinction debt.

Habitat destruction can also decrease the range of certain organism populations. This can result in the reduction of genetic diversity and perhaps the production of infertile youths, as these organisms would have a higher possibility of mating with related organisms within their population, or different species.

One of the most famous examples is the impact upon China's Giant Panda, once found across the nation. Now it is only found in fragmented and isolated regions in the south-west of the country, as a result of widespread deforestation in the 20th Century.

Geography

Biodiversity hotspots are chiefly tropical regions that feature high concentrations of endemic species and, when all hotspots are combined, may contain over half of the world's terrestrial species. These hotspots are suffering from habitat loss and destruction. Most of the natural habitat on islands and in areas of high human population density has already been destroyed (WRI, 2003).

Islands suffering extreme habitat destruction include New Zealand, Madagascar, the Philippines, and Japan. South and east Asia—especially China, India, Malaysia, Indonesia, and Japan—and many areas in West Africa have extremely dense human populations that allow little room for natural habitat. Marine areas close to highly populated coastal cities also face degradation of their coral reefs or other marine habitat. These areas include the eastern coasts of Asia and Africa, northern coasts of South America, and the Caribbean Sea and its associated islands.

Regions of unsustainable agriculture or unstable governments, which may go hand-in-hand, typically experience high rates of habitat destruction. Central America, Sub-Saharan Africa, and the Amazonian tropical rainforest areas of South America are the main regions with unsustainable agricultural practices or government mismanagement.

Areas of high agricultural output tend to have the highest extent of habitat destruction. In the U.S., less than 25% of native vegetation remains in many parts of the East and Midwest. Only 15% of land area remains unmodified by human activities in all of Europe.

Ecosystems

Tropical rainforests have received most of the attention concerning the destruction of habitat. From the approximately 16 million square kilometres of tropical rainforest habitat that originally existed worldwide, less than 9 million square kilometres remain today. The current rate of deforestation is 160,000 square kilometres per year, which equates to a loss of approximately 1% of original forest habitat each year.

Other forest ecosystems have suffered as much or more destruction as tropical rainforests. Farming and logging have severely disturbed at least 94% of temperate broadleaf forests; many old growth forest stands have lost more than 98% of their previous area because of human activities. Tropical deciduous dry forests are easier to clear and burn and are more suitable for agriculture and cattle ranching than tropical rainforests; consequently, less than 0.1% of dry forests in Central America's Pacific Coast and less than 8% in Madagascar remain from their original extents.

Plains and desert areas have been degraded to a lesser extent. Only 10-20% of the world's drylands, which include temperate grasslands, savannas, and shrublands, scrub and deciduous forests, have been somewhat degraded. But included in that 10-20% of land is the approximately 9 million square kilometres of seasonally drylands that humans have converted to deserts through the process of desertification. The tallgrass prairies of North America, on the other hand, have less than 3% of natural habitat remaining that has not been converted to farmland.

Wetlands and marine areas have endured high levels of habitat destruction. More than 50% of wetlands in the U.S. have been destroyed in just the last 200 years. Between 60% and 70% of European wetlands have been completely destroyed. About one-fifth (20%) of marine coastal areas have been highly modified by humans. One-fifth of coral reefs have also been destroyed, and another fifth has been severely degraded by overfishing, pollution, and invasive species; 90% of the Philippines' coral reefs alone have been destroyed. Finally, over 35% mangrove ecosystems worldwide have been destroyed.

Human Causes

Habitat destruction caused by humans includes conversion of land to agriculture, urban sprawl, infrastructure development, and other anthropogenic changes to the characteristics of land. Habitat degradation, fragmentation, and pollution are aspects of habitat destruction caused by humans that do not necessarily involve overt destruction of habitat, yet result in habitat collapse. Desertification, deforestation, and coral reef degradation are specific types of habitat destruction for those areas (deserts, forests, coral reefs).

Geist and Lambin (2002) assessed 152 case studies of net losses of tropical forest cover to determine any patterns in the proximate and underlying causes of tropical deforestation. Their results, yielded as percentages of the case studies in which each parameter was a significant factor, provide a quantitative prioritization of which proximate and underlying causes were the most significant. The proximate causes were clustered into broad categories of agricultural expansion (96%), infrastructure expansion (72%), and wood extraction (67%).

Therefore, according to this study, forest conversion to agriculture is the main land use change responsible for tropical deforestation.

The specific categories reveal further insight into the specific causes of tropical deforestation: transport extension (64%), commercial wood extraction (52%), permanent cultivation (48%), cattle ranching (46%), shifting (slash and burn) cultivation (41%), subsistence agriculture (40%), and fuel wood extraction for domestic use (28%). One result is that shifting cultivation is not the primary cause of deforestation in all world regions, while transport extension (including the construction of new roads) is the largest single proximate factor responsible for deforestation.

Drivers

While the above-mentioned activities are the proximal or direct causes of habitat destruction in that they actually destroy habitat, this still does not identify why humans destroy habitat. The forces that cause humans to destroy habitat are known as *drivers* of habitat destruction. Demographic, economic, sociopolitical, scientific and technological, and cultural drivers all contribute to habitat destruction.

Demographic drivers include the expanding human population; rate of population increase over time; spatial distribution of people in a given area (urban versus rural), ecosystem type, and country; and the combined effects of poverty, age, family planning, gender, and education status of people in certain areas. Most of the exponential human population growth worldwide is occurring in or close to biodiversity hotspots.

This may explain why human population density accounts for 87.9% of the variation in numbers of threatened species across 114 countries, providing indisputable evidence that people play the largest role in decreasing biodiversity.

The boom in human population and migration of people into such species-rich regions are making conservation efforts not only more urgent but also more likely to conflict with local human interests. The high local population density in such areas is directly correlated to the poverty status of the local people, most of whom lack an education and family planning.

From the Geist and Lambin (2002) study described in the previous section, the underlying driving forces were prioritized as follows (with the percent of the 152 cases the factor played a significant role in): economic factors (81%), institutional or policy factors (78%), technological factors (70%), cultural or socio-political factors (66%), and demographic factors (61%).

The main economic factors included commercialization and growth of timber markets (68%), which are driven by national and international demands; urban industrial growth (38%); low domestic costs for land, labour, fuel, and timber (32%); and increases in product prices mainly for cash crops (25%). Institutional and policy factors included formal pro-deforestation policies on land development (40%), economic growth including colonization and infrastructure improvement (34%), and subsidies for land-based activities (26%); property rights and land-tenure insecurity (44%); and policy failures such as corruption, lawlessness, or mismanagement (42%).

The main technological factor was the poor application of technology in the wood industry (45%), which leads to wasteful logging practices. Within the broad category of cultural and sociopolitical factors are public attitudes and values (63%), individual /household behaviour (53%), public unconcern toward forest environments (43%), missing basic values (36%), and unconcern by individuals (32%). Demographic factors were the in-migration of colonizing settlers into sparsely populated forest areas (38%) and growing population density—a result of the first factor—in those areas (25%).

There are also feedbacks and interactions among the proximate and underlying causes of deforestation that can amplify the process. Road construction has the largest feedback effect, because it interacts with—and leads to—the establishment of new settlements and more people, which causes a growth in wood (logging) and food markets.

Growth in these markets, in turn, progresses the commercialization of agriculture and logging industries. When these industries become commercialized, they must become more efficient by utilising larger or more modern machinery that often are worse on the habitat than traditional farming and logging methods.

Either way, more land is cleared more rapidly for commercial markets. This common feedback example manifests just how closely related the proximate and underlying causes are to each other.

Conservation and Management of Ecosystems

Ecosystem management is a process that aims to conserve major ecological services and restore natural resources while meeting the socioeconomic, political and cultural needs of current and future generations.

The principal objective of ecosystem management is the efficient maintenance, and ethical use of natural resources. Ecosystem management acknowledges that the interrelation of socio-cultural, economic and ecological systems is paramount to understanding the circumstances that affect environmental goals and outcomes.

It is a multifaceted and holistic approach which requires a significant change in how the natural and human environments are identified.

Several approaches to effective ecosystem management engage conservation efforts at both a local or landscape level and involves: adaptive management, natural resource management, strategic management, and command and control management.

Stakeholders

Stakeholders are individuals or groups of people who are affected by environmental decisions and actions, but they also may have power to influence the outcomes of environmental decisions relating to ecosystem management (Reed et al. 2009).

The complex nature of decisions made in ecosystem management, from local to international scales, requires stakeholder participation from a diversity of knowledge, perceptions and values of nature (Billgren and Holmen 2007; and Reed 2008).

Stakeholders will often have different interests in ecosystem services (Shepherd 2008) which means effective management of ecosystems requires a negotiation process that develops mutual trust in issues of common interest with the objective of creating mutually beneficial partnerships (Mushove and Vogel 2005).

Adaptive Management

Adaptive management is based on the concept that predicting future influences/disturbance to an ecosystem is limited and unclear (Pahl-Wostl 2007).

Therefore, the goal of adaptive management is to manage the ecosystem so it maintains the greatest amount of ecological integrity, but also to utilise management practices that have the ability to change based on new experience and insights (Department of Interior, Holling 1978, Pahl-Wostl 2007).

Adaptive management aims to identify uncertainties in the management of an ecosystem while using hypothesis testing to further understand the system (Resilience Alliance 2010). In this regard, adaptive management encourages learning from the outcomes of previously implemented management strategies (Pahl-Wostl 2007). Ecosystem managers form hypotheses about the ecosystem and its functionality and then implement different management techniques to test the hypotheses (DOI). The implemented techniques are then analyzed to evaluate any regressions or improvements in functionality of the ecosystem caused by the technique (DOI). Further analysis allows for modification of the technique until it successfully meets the ecological needs of the ecosystem (Holling 1978). Thus, adaptive management serves as a "learning by doing" method for ecosystem management.

Adaptive management has had mixed success in the field of ecosystem management (Gregory et al. 2006). This is because ecosystem managers may not be equipped with the decision-making skills needed to undertake an adaptive management methodology (Gregory et al. 2006).

Additionally, economic, social and political priorities can interfere with adaptive management decisions (Gregory et al. 2006). For this reason, adaptive management should be a social process as well as scientific, focusing on institutional strategies while implementing experimental management techniques (Resilience Alliance 2010).

Natural Resource Management

The term natural resource management is frequently used when dealing with a particular resource for human use rather than managing the whole ecosystem (Kellert et al. 2000). A main objective of natural resources management is the sustainability for future generations, which appoints ecosystem managers to balance natural resources exploitation and conservation over long-term timeframe (Ascher 2001).

The balanced relationship of each resource in an ecosystem is subject to change at different spatial and temporal scales (Boyce & Haney 1997). Dimensions such as, watersheds, soils, flora and fauna, need to be considered individually and on a landscape level. A variety of natural resources are utilised for food, medicine, energy and shelter (Chapin, Kofinas & Folke 2009).

The ecosystem management concept is based on the relationship between sustainable resource maintenance and human demand for use of natural resources (Boyce & Heney, 1997). Therefore, socioeconomics factors significantly affect natural resource management (Kellert et al. 2000).

The goal of a natural resource manager is to fulfill the demand for a given resource without causing harm to the ecosystem, or jeopardizing the future of the resource (Grimble & Wellard 1997). Partnerships between ecosystem managers, natural resource managers and stakeholders should be encouraged in order to promote a more sustainable use of limited natural resources (Cork, Stoneham & Lowe 2007). Natural resource managers must initially measure the overall integrity of the ecosystem they are involved in. If the ecosystem supporting resources is healthy, managers can decide on the ideal amount of resource extraction, while leaving enough to allow the resource to replenish itself for subsequent harvests (Grimble & Wellard 1997).

Historically, some natural resources have experienced limited human disturbance and therefore have been able to subsist naturally. However, some ecosystems such as forests, which typically provide considerable timber resources; have sometimes undergone successful reforestation processes and consequently have accommodated the needs of future generations.

A successfully managed resource, will provide for current demand while leaving enough to repopulate and provide for future demand. Human populations have been increasing rapidly, introducing new stressors to ecosystems, such as climate change and influxes of invasive species. As a result, the demand for natural resources is unpredictable (Chapin, Kofinas & Folke 2009). Although ecosystem changes may occur gradually, the cumulative changes can have negative effects for humans and wildlife (Ascher 2001). Geographic Information Systems (GIS) and Remote Sensing Applications can be used to monitor and evaluate natural resources by mapping them in local and global scales. These tools will continue to be highly beneficial in natural resources management (Boyce & Haney 1997).

Strategic Management

Strategic management encourages the establishment of goals that will benefit the ecosystem while keeping socioeconomic and politically relevant issues in mind (Brussard et al. 1998). Strategic management differs from other types of ecosystem management because it keeps stakeholders involved and relies on their input to develop the best management strategy for an ecosystem. Similarly to other modes of ecosystem management, this method places a high level of importance on evaluating and reviewing any changes, progress or negative impacts and prioritizes flexibility in adapting management protocols as a result of new information (Shmelev and Powell 2006).

Landscape Level Conservation

Landscape level conservation is a method that considers wildlife needs at a broader landscape level scale when implementing conservation initiatives (AWF). This approach to ecosystem management involves the consideration of broad scale interconnected ecological systems that acknowledges the whole scope of an environmental problem (Boyce & Haney 1997). In a human–dominated world, weighing the landcape requirements of wildlife versus the needs of humans is a complicated matter (Opdam & Wascher 2003). Landscape level conservation is carried out in a number of ways. Wildlife corridors, for example, are connections between otherwise isolated habitat patches that are proposed as a solution to habitat fragmentation (Hudgens & Haddad 2003).

In some landscape level conservation approaches, a key species vulnerable to landscape alteration is identified and its habitat requirements are assessed in order to identify the best option for protecting their ecosystem (Lambeck 1997). However, lining up the habitat requirements of numerous species in an ecosystem can be difficult, which is why more comprehensive approaches to further understand these variations have been considered in landscape level conservation (Vos 2010). Human-induced environmental degradation is an increasing problem globally, which is why landscape level ecology plays an important role in ecosystem management (Velazquez et al. 2003). Traditional conservation methods targeted at individual species need to be modified to include the maintenance of wildlife habitats through consideration of both human-induced and natural environmental factors (Velazquez et al. 2003).

Command and Control Management

Command and control management utilises a linear problem solving approach where a perceived problem is solved through controlling devices such as laws, threats, contracts and/or agreements (Holling & Meffe 1996).

This top-down approach is used across many disciplines and works best with problems that are relatively simple, well-defined and work in terms of cause and effect (Knight & Meffe 1997).

The application of command and control management has often attempted to control nature in order to improve product extractions, establish predictability and reduce threats (Holling & Meffe 1996). Some obvious examples of command and control management actions include: the use of herbicides and pesticides to safeguard crops in order to harvest more products; the culling of predators in order to obtain larger, more reliable game species; and the safeguarding of timber supply, by suppressing forest fires.

Attempts at command and control management often backfire (a literal problem in forests that have been 'protected' from fire by humans and are subsequently full of fuel build-up) in ecosystems due to their inherent complexities. Consequently there has been a transition away from command and control management due to many undesirable outcomes and a stronger focus has been placed on more holistic approaches that focus on adaptive management and finding solutions through partnerships (Knight & Meffe 1997).

Environmental Hazards and Problems of Pollution

Pollution is the introduction of contaminants into a natural environment that causes instability, disorder, harm or discomfort to the ecosystem i.e. physical systems or living organisms. Pollution can take the form of chemical substances or energy, such as noise, heat, or light. Pollutants, the elements of pollution, can be foreign substances or energies, or naturally occurring; when naturally occurring, they are considered contaminants when they exceed natural levels. Pollution is often classed as point source or nonpoint source pollution. The Blacksmith Institute issues an annual list of the world's worst polluted places. In the 2007 issues the ten top nominees are located in Azerbaijan, China, India, Peru, Russia, Ukraine, and Zambia.

Ancient Cultures

Air pollution has always been with us. According to a 1983 article in the journal *Science,* "soot found on ceilings of prehistoric caves provides ample evidence of the high levels of pollution that was associated with inadequate ventilation of open fires." The forging of metals appears to be a key turning point in the creation of significant air pollution levels outside the home. Core samples of glaciers in Greenland indicate increases in pollution associated with Greek, Roman and Chinese metal production.

Official Acknowledgement

King Edward I of England banned the burning of sea-coal by proclamation in London in 1272, after its smoke had become a problem. But the fuel was so common in England that this earliest of names for it was acquired because it could be carted away from some shores by the wheelbarrow. Air pollution would continue to be a problem in England, especially later during the industrial revolution, and extending into the recent past with the Great Smog of 1952. London also recorded one of the earlier extreme cases of water quality problems with the Great Stink on the Thames of 1858, which led to construction of the London sewerage system soon afterward.

It was the industrial revolution that gave birth to environmental pollution as we know it today. The emergence of great factories and consumption of immense quantities of coal and other fossil fuels gave rise to unprecedented air pollution and the large volume of industrial chemical discharges added to the growing load of untreated human waste.

Chicago and Cincinnati were the first two American cities to enact laws ensuring cleaner air in 1881.

Other cities followed around the country until early in the 20th century, when the short lived Office of Air Pollution was created under the Department of the Interior. Extreme smog events were experienced by the cities of Los Angeles and Donora, Pennsylvania in the late 1940s, serving as another public reminder.

Modern Awareness

Pollution became a popular issue after World War II, due to radioactive fallout from atomic warfare and testing. Then a non-nuclear event, The Great Smog of 1952 in London, killed at least 4000 people. This prompted some of the first major modern environmental legislation, The Clean Air Act of 1956. Pollution began to draw major public attention in the United States between the mid-1950s and early 1970s, when Congress passed the Noise Control Act, the Clean Air Act, the Clean Water Act and the National Environmental Policy Act.

Bad bouts of local pollution helped increase consciousness. PCB dumping in the Hudson River resulted in a ban by the EPA on consumption of its fish in 1974.

Long-term dioxin contamination at Love Canal starting in 1947 became a national news story in 1978 and led to the Superfund legislation of 1980. Legal proceedings in the 1990s helped bring to light Chromium-6 releases in California—the champions of whose victims became famous. The pollution of industrial land gave rise to the name brownfield, a term now common in city planning. DDT was banned in most of the developed world after the publication of Rachel Carson's *Silent Spring*.

The development of nuclear science introduced radioactive contamination, which can remain lethally radioactive for hundreds of thousands of years.

Lake Karachay, named by the Worldwatch Institute as the "most polluted spot" on earth, served as a disposal site for the Soviet Union thoroughout the 1950s and 1960s. Second place may go to the area of Chelyabinsk U.S.S.R. as the "Most polluted place on the planet".

Nuclear weapons continued to be tested in the Cold War, sometimes near inhabited areas, especially in the earlier stages of their development.

The toll on the worst-affected populations and the growth since then in understanding about the critical threat to human health posed by radioactivity has also been a prohibitive complication associated with nuclear power.

Though extreme care is practiced in that industry, the potential for disaster suggested by incidents such as those at Three Mile Island and Chernobyl pose a lingering specter of public mistrust. One legacy of nuclear testing before most forms were banned has been significantly raised levels of background radiation.

International catastrophes such as the wreck of the Amoco Cadiz oil tanker off the coast of Brittany in 1978 and the Bhopal disaster in 1984 have demonstrated the universality of such events and the scale on which efforts to address them needed to engage.

The borderless nature of atmosphere and oceans inevitably resulted in the implication of pollution on a planetary level with the issue of global warming. Most recently the term persistent organic pollutant (POP) has come to describe a group of chemicals such as PBDEs and PFCs among others.

Though their effects remain somewhat less well understood owing to a lack of experimental data, they have been detected in various ecological habitats far removed from industrial activity such as the Arctic, demonstrating diffusion and bioaccumulation after only a relatively brief period of widespread use.

Growing evidence of local and global pollution and an increasingly informed public over time have given rise to environmentalism and the environmental movement, which generally seek to limit human impact on the environment.

Forms of Pollution

The major forms of pollution are listed below along with the particular pollutants relevant to each of them:

- Air pollution, the release of chemicals and particulates into the atmosphere. Common gaseous pollutants include carbon monoxide, sulfur dioxide, chlorofluorocarbons (CFCs) and nitrogen oxides produced by industry and motor vehicles.

  Photochemical ozone and smog are created as nitrogen oxides and hydrocarbons react to sunlight. Particulate matter, or fine dust is characterised by their micrometre size PM_{10} to $PM_{2.5}$.
- Light pollution, includes light trespass, over-illumination and astronomical interference.
- Littering
- Noise pollution, which encompasses roadway noise, aircraft noise, industrial noise as well as high-intensity sonar.
- Soil contamination occurs when chemicals are released intentionally, by spill or underground leakage.

  Among the most significant soil contaminants are hydrocarbons, heavy metals, MTBE, herbicides, pesticides and chlorinated hydrocarbons.
- Radioactive contamination, resulting from 20th century activities in atomic physics, such as nuclear power generation and nuclear weapons research, manufacture and deployment.
- Thermal pollution, is a temperature change in natural water bodies caused by human influence, such as use of water as coolant in a power plant.
- Visual pollution, which can refer to the presence of overhead power lines, motorway billboards, scarred landforms (as from strip mining), open storage of trash or municipal solid waste.

- Water pollution, by the discharge of wastewater from commercial and industrial waste (intentionally or through spills) into surface waters; discharges of untreated domestic sewage, and chemical contaminants, such as chlorine, from treated sewage; release of waste and contaminants into surface runoff flowing to surface waters (including urban runoff and agricultural runoff, which may contain chemical fertilizers and pesticides); waste disposal and leaching into groundwater; eutrophication and littering.

Pollutants

A pollutant is a waste material that pollutes air, water or soil. Three factors determine the severity of a pollutant: its chemical nature, the concentration and the persistence.

Sources and Causes

Air pollution comes from both natural and man made sources. Though globally man made pollutants from combustion, construction, mining, agriculture and warfare are increasingly significant in the air pollution equation.

Motor vehicle emissions are one of the leading causes of air pollution. China, United States, Russia, Mexico, and Japan are the world leaders in air pollution emissions. Principal stationary pollution sources include chemical plants, coal-fired power plants, oil refineries, petrochemical plants, nuclear waste disposal activity, incinerators, large livestock farms (dairy cows, pigs, poultry, etc.), PVC factories, metals production factories, plastics factories, and other heavy industry. Agricultural air pollution comes from contemporary practices which include clear felling and burning of natural vegetation as well as spraying of pesticides and herbicides.

About 400 million metric tons of hazardous wastes are generated each year. The United States alone produces about 250 million metric tons. Americans constitute less than 5% of the world's population, but produce roughly 25% of the world's CO_2, and generate approximately 30% of world's waste. In 2007, China has overtaken the United States as the world's biggest producer of CO_2, while still far behind based on per capita pollution - ranked 78th among the world's nation.

In February 2007, a report by the Intergovernmental Panel on Climate Change (IPCC), representing the work of 2,500 scientists, economists, and policymakers from more than 120 countries, said that humans have been the primary cause of global warming since 1950.

Humans have ways to cut greenhouse gas emissions and avoid the consequences of global warming, a major climate report concluded. But in order to change the climate, the transition from fossil fuels like coal and oil needs to occur within decades, according to the final report this year from the UN's Intergovernmental Panel on Climate Change (IPCC).

Some of the more common soil contaminants are chlorinated hydrocarbons (CFH), heavy metals (such as chromium, cadmium–found in rechargeable batteries, and lead–found in lead paint, aviation fuel and still in some countries, gasoline), MTBE, zinc, arsenic and benzene.

In 2001 a series of press reports culminating in a book called *Fateful Harvest* unveiled a widespread practice of recycling industrial byproducts into fertilizer, resulting in the contamination of the soil with various metals.

Ordinary municipal landfills are the source of many chemical substances entering the soil environment (and often groundwater), emanating from the wide variety of refuse accepted, especially substances illegally discarded there, or from pre-1970 landfills that may have been subject to little control in the U.S. or EU. There have also been some unusual releases of polychlorinated dibenzodioxins, commonly called *dioxins* for simplicity, such as TCDD.

Pollution can also be the consequence of a natural disaster. For example, hurricanes often involve water contamination from sewage, and petrochemical spills from ruptured boats or automobiles. Larger scale and environmental damage is not uncommon when coastal oil rigs or refineries are involved.

Some sources of pollution, such as nuclear power plants or oil tankers, can produce widespread and potentially hazardous releases when accidents occur.

In the case of noise pollution the dominant source class is the motor vehicle, producing about ninety percent of all unwanted noise worldwide.

Effects

Human Health

Adverse air quality can kill many organisms including humans. Ozone pollution can cause respiratory disease, cardiovascular disease, throat inflammation, chest pain, and congestion. Water pollution causes approximately 14,000 deaths per day, mostly due to contamination of drinking water by untreated sewage in developing countries. An estimated 700 million Indians have no access to a proper toilet, and 1,000 Indian children die of diarrhoeal sickness every day. Nearly 500 million Chinese lack access to safe drinking water. 656,000 people die prematurely each year in China because of air pollution. In India, air pollution is believed to cause 527,700 fatalities a year. Studies have estimated that the number of people killed annually in the US could be over 50,000.

Oil spills can cause skin irritations and rashes. Noise pollution induces hearing loss, high blood pressure, stress, and sleep disturbance. Mercury has been linked to developmental deficits in children and neurologic symptoms. Older people are majorly exposed to diseases induced by air pollution. Those with heart or lung disorders are under additional risk. Children and infants are also at serious risk. Lead and other heavy metals have been shown to cause neurological problems. Chemical and radioactive substances can cause cancer and as well as birth defects.

Environment

Pollution has been found to be present widely in the environment. There are a number of effects of this:

- Biomagnification describes situations where toxins (such as heavy metals) may pass through trophic levels, becoming exponentially more concentrated in the process.
- Carbon dioxide emissions cause ocean acidification, the ongoing decrease in the pH of the Earth's oceans as CO_2 becomes dissolved.
- The emission of greenhouse gases leads to global warming which affects ecosystems in many ways.
- Invasive species can out compete native species and reduce biodiversity. Invasive plants can contribute debris and biomolecules (allelopathy) that can alter soil and chemical compositions of an environment, often reducing native species competitiveness.
- Nitrogen oxides are removed from the air by rain and fertilise land which can change the species composition of ecosystems.
- Smog and haze can reduce the amount of sunlight received by plants to carry out photosynthesis and leads to the production of tropospheric ozone which damages plants.

- Soil can become infertile and unsuitable for plants. This will affect other organisms in the food web.
- Sulphur dioxide and nitrogen oxides can cause acid rain which lowers the pH value of soil.

Environmental Health Information

The Toxicology and Environmental Health Information Program (TEHIP) at the United States National Library of Medicine (NLM) maintains a comprehensive toxicology and environmental health web site that includes access to resources produced by TEHIP and by other government agencies and organisations.

This web site includes links to databases, bibliographies, tutorials, and other scientific and consumer-oriented resources. TEHIP also is responsible for the Toxicology Data Network (TOXNET®) an integrated system of toxicology and environmental health databases that are available free of charge on the web.

TOXMAP is a Geographic Information System (GIS) that is part of TOXNET.

TOXMAP uses maps of the United States to help users visually explore data from the United States Environmental Protection Agency's (EPA) Toxics Release Inventory and Superfund Basic Research Programs.

Regulation and Monitoring

To protect the environment from the adverse effects of pollution, many nations worldwide have enacted legislation to regulate various types of pollution as well as to mitigate the adverse effects of pollution.

Pollution Control

Pollution control is a term used in environmental management. It means the control of emissions and effluents into air, water or soil. Without pollution control, the waste products from consumption, heating, agriculture, mining, manufacturing, transportation and other human activities, whether they accumulate or disperse, will degrade the environment. In the hierarchy of controls, pollution prevention and waste minimisation are more desirable than pollution control. In the field of land development, low impact development is a similar technique for the prevention of urban runoff.

Practices

- recycling

Pollution Control Devices

- Dust collection systems
  - Baghouses
  - Cyclones
  - Electrostatic precipitators
- Scrubbers
  - Baffle spray scrubber
  - Cyclonic spray scrubber
  - Ejector venturi scrubber
  - Mechanically aided scrubber
  - Spray tower
  - Wet scrubber
- Sewage treatment
  - Sedimentation (Primary treatment)
  - Activated sludge biotreaters (Secondary treatment; also used for industrial wastewater)
  - Aerated lagoons
  - Constructed wetlands (also used for urban runoff)
- Industrial wastewater treatment
  - API oil-water separators
  - Biofilters
  - Dissolved air flotation (DAF)
  - Powdered activated carbon treatment
  - Ultrafiltration
- Vapor recovery systems

Perspectives

The earliest precursor of pollution generated by life forms would have been a natural function of their existence.

The attendant consequences on viability and population levels fell within the sphere of natural selection.

These would have included the demise of a population locally or ultimately, species extinction. Processes that were untenable would have resulted in a new balance brought about by changes and adaptations. At the extremes, for any form of life, consideration of pollution is superseded by that of survival.

For humankind, the factor of technology is a distinguishing and critical consideration, both as an enabler and an additional source of byproducts.

Short of survival, human concerns include the range from quality of life to health hazards. Since science holds experimental demonstration to be definitive, modern treatment of toxicity or environmental harm involves defining a level at which an effect is observable.

Common examples of fields where practical measurement is crucial include automobile emissions control, industrial exposure (e.g. Occupational Safety and Health Administration (OSHA) PELs), toxicology (e.g. LD_{50}), and medicine (e.g. medication and radiation doses).

"The solution to pollution is dilution", is a dictum which summarises a traditional approach to pollution management whereby sufficiently diluted pollution is not harmful. It is well-suited to some other modern, locally scoped applications such as laboratory safety procedure and hazardous material release emergency management. But it assumes that the dilutant is in virtually unlimited supply for the application or that resulting dilutions are acceptable in all cases.

Such simple treatment for environmental pollution on a wider scale might have had greater merit in earlier centuries when physical survival was often the highest imperative, human population and densities were lower, technologies were simpler and their byproducts more benign. But these are often no longer the case. Furthermore, advances have enabled measurement of concentrations not possible before. The use of statistical methods in evaluating outcomes has given currency to the principle of probable harm in cases where assessment is warranted but resorting to deterministic models is impractical or unfeasible. In addition, consideration of the environment beyond direct impact on human beings has gained prominence.

Yet in the absence of a superseding principle, this older approach predominates practices throughout the world. It is the basis by which to gauge concentrations of effluent for legal release, exceeding which penalties are assessed or restrictions applied. The regressive cases are those where a controlled level of release is too high or, if enforceable, is neglected. Migration from pollution dilution to elimination in many cases is confronted by challenging economical and technological barriers.

Greenhouse Gases and Global Warming

Carbon dioxide, while vital for photosynthesis, is sometimes referred to as pollution, because raised levels of the gas in the atmosphere are affecting the Earth's climate. Disruption of the environment can also highlight the connection between areas of pollution that would normally be classified separately, such as those of water and air.

Recent studies have investigated the potential for long-term rising levels of atmospheric carbon dioxide to cause slight but critical increases in the acidity of ocean waters, and the possible effects of this on marine ecosystems.

Ozone Depletion

Ozone depletion describes two distinct but related phenomena observed since the late 1970s: a steady decline of about 4% per decade in the total volume of ozone in Earth's stratosphere (the ozone layer), and a much larger springtime decrease in stratospheric ozone over Earth's polar regions. The latter phenomenon is referred to as the ozone hole. In addition to these well-known stratospheric phenomena, there are also springtime polar tropospheric ozone depletion events.

The details of polar ozone hole formation differ from that of mid-latitude thinning, but the most important process in both is catalytic destruction of ozone by atomic halogens. The main source of these halogen atoms in the stratosphere is photodissociation of man-made Halocarbon refrigerants (CFCs, freons, halons.) These compounds are transported into the stratosphere after being emitted at the surface. Both types of ozone depletion were observed to increase as emissions of Halocarbons increased.

CFCs and other contributory substances are referred to as ozone-depleting substances (ODS). Since the ozone layer prevents most harmful UVB wavelengths (280–315 nm) of ultraviolet light (UV light) from passing through the Earth's atmosphere, observed and projected decreases in ozone have generated worldwide concern leading to adoption of the Montreal Protocol that bans the production of CFCs, halons as and other ozone-depleting chemicals such as carbon tetrachloride and trichloroethane.

It is suspected that a variety of biological consequences such as increases in skin cancer, cataracts, damage to plants, and reduction of plankton populations in the ocean's photic zone may result from the increased UV exposure due to ozone depletion.

Ozone Cycle Overview

Three forms (or allotropes) of oxygen are involved in the ozone-oxygen cycle: oxygen atoms (O or atomic oxygen), oxygen gas (O_2 or diatomic oxygen), and ozone gas (O_3 or triatomic oxygen). Ozone is formed in the stratosphere when oxygen molecules photodissociate after absorbing an ultraviolet photon whose wavelength is shorter than 240 nm.

This converts a single O_2 into two atomic oxygen ions. The atomic oxygen ions then combine with separate O_2 molecules to create two O_3 molecules.

These ozone molecules absorb UV light between 310 and 200 nm, following which ozone splits into a molecule of O_2 and an oxygen atom. The oxygen atom then joins up with an oxygen molecule to regenerate ozone. This is a continuing process which terminates when an oxygen atom "recombines" with an ozone molecule to make two O_2 molecules.

$$O + O_3 \rightarrow 2\ O_2$$

The overall amount of ozone in the stratosphere is determined by a balance between photochemical production and recombination.

Ozone can be destroyed by a number of free radical catalysts, the most important of which are the hydroxyl radical (OH ·), the nitric oxide radical (NO ·), the atomic chlorine ion (Cl ·) and the atomic bromine ion (Br ·).

All of these have both natural and man-made sources; at the present time, most of the OH · and NO · in the stratosphere is of natural origin, but human activity has dramatically increased the levels of chlorine and bromine.

These elements are found in certain stable organic compounds, especially chlorofluorocarbons (CFCs), which may find their way to the stratosphere without being destroyed in the troposphere due to their low reactivity. Once in the stratosphere, the Cl and Br atoms are liberated from the parent compounds by the action of ultraviolet light, e.g.

$$CFCl_3 + \text{electromagnetic radiation} \rightarrow CFCl_2 + Cl$$

The Cl and Br atoms can then destroy ozone molecules through a variety of catalytic cycles.

In the simplest example of such a cycle, a chlorine atom reacts with an ozone molecule, taking an oxygen atom with it (forming ClO) and leaving a normal oxygen molecule.

The chlorine monoxide (i.e., the ClO) can react with a second molecule of ozone (i.e., O_3) to yield another chlorine atom and two molecules of oxygen. The chemical shorthand for these gas-phase reactions is:

- $Cl + O_3 \rightarrow ClO + O_2$ - The chlorine atom changes an ozone molecule to ordinary oxygen.
- $ClO + O_3 \rightarrow Cl + 2\ O_2$ - The ClO from the previous reaction destroys a second ozone molecule and recreates the original chlorine atom, which can repeat the first reaction and continue to destroy ozone

The overall effect is a decrease in the amount of ozone. More complicated mechanisms have been discovered that lead to ozone destruction in the lower stratosphere as well.

A single chlorine atom would keep on destroying ozone (thus a catalyst) for up to two years (the time scale for transport back down to the troposphere) were it not for reactions that remove them from this cycle by forming reservoir species such as hydrogen chloride (HCl) and chlorine nitrate ($ClONO_2$).

On a per atom basis, bromine is even more efficient than chlorine at destroying ozone, but there is much less bromine in the atmosphere at present.

As a result, both chlorine and bromine contribute significantly to the overall ozone depletion. Laboratory studies have shown that fluorine and iodine atoms participate in analogous catalytic cycles. However, in the Earth's stratosphere, fluorine atoms react rapidly with water and methane to form strongly bound HF, while organic molecules which contain iodine react so rapidly in the lower atmosphere that they do not reach the stratosphere in significant quantities. Furthermore, a single chlorine atom is able to react with 100,000 ozone molecules. This fact plus the amount of chlorine released into the atmosphere by chlorofluorocarbons (CFCs) yearly demonstrates how dangerous CFCs are to the environment.

Observations on Ozone Layer Depletion

The most pronounced decrease in ozone has been in the lower stratosphere. However, the ozone hole is most usually measured not in terms of ozone concentrations at these levels (which are typically of a few parts per million) but by reduction in the total *column ozone*, above a point on the Earth's surface, which is normally expressed in Dobson units, abbreviated as "DU".

Marked decreases in column ozone in the Antarctic spring and early summer compared to the early 1970s and before have been observed using instruments such as the Total Ozone Mapping Spectrometre (TOMS).

Reductions of up to 70% in the ozone column observed in the austral (southern hemispheric) spring over Antarctica and first reported in 1985 (Farman et al. 1985) are continuing.

Through the 1990s, total column ozone in September and October have continued to be 40–50% lower than pre-ozone-hole values. In the Arctic the amount lost is more variable year-to-year than in the Antarctic. The greatest declines, up to 30%, are in the winter and spring, when the stratosphere is colder.

Reactions that take place on polar stratospheric clouds (PSCs) play an important role in enhancing ozone depletion. PSCs form more readily in the extreme cold of Antarctic stratosphere. This is why ozone holes first formed, and are deeper, over Antarctica. Early models failed to take PSCs into account and predicted a gradual global depletion, which is why the sudden Antarctic ozone hole was such a surprise to many scientists.In middle latitudes it is preferable to speak of ozone depletion rather than holes. Declines are about 3% below pre-1980 values for 35–60°N and about 6% for 35–60°S. In the tropics, there are no significant trends.

Ozone depletion also explains much of the observed reduction in stratospheric and upper tropospheric temperatures. The source of the warmth of the stratosphere is the absorption of UV radiation by ozone, hence reduced ozone leads to cooling. Some stratospheric cooling is also predicted from increases in greenhouse gases such as CO_2; however the ozone-induced cooling appears to be dominant.

Predictions of ozone levels remain difficult. The World Meteorological Organisation Global Ozone Research and Monitoring Project—Report No. 44 comes out strongly in favour for the Montreal Protocol, but notes that a UNEP 1994 Assessment overestimated ozone loss for the 1994–1997 period.

Chemicals in the Atmosphere

CFCs and Related Compounds in the Atmosphere

Chlorofluorocarbons (CFCs) and other halogenated ozone depleting substances (ODS) are mainly responsible for man-made chemical ozone depletion.

The total amount of effective halogens (chlorine and bromine) in the stratosphere can be calculated and are known as the equivalent effective stratospheric chlorine (EESC).

CFCs were invented by Thomas Midgley, Jr. in the 1920s. They were used in air conditioning/cooling units, as aerosol spray propellants prior to the 1980s, and in the cleaning processes of delicate electronic equipment. They also occur as by-products of some chemical processes. No significant natural sources have ever been identified for these compounds — their presence in the atmosphere is due almost entirely to human manufacture. As mentioned in the *ozone cycle overview* above, when such ozone-depleting chemicals reach the stratosphere, they are dissociated by ultraviolet light to release chlorine atoms. The chlorine atoms act as a catalyst, and each can break down tens of thousands of ozone molecules before being removed from the stratosphere. Given the longevity of CFC molecules, recovery times are measured in decades. It is calculated that a CFC molecule takes an average of about five to seven years to go from the ground level up to the upper atmosphere, and it can stay there for about a century, destroying up to one hundred thousand ozone molecules during that time.

Verification of Observations

Scientists have been increasingly able to attribute the observed ozone depletion to the increase of man-made (anthropogenic) halogen compounds from CFCs by the use of complex chemistry transport models and their validation against observational data (e.g. SLIMCAT, CLaMS — Chemical Lagrangian Model of the Stratosphere). These models work by combining satellite measurements of chemical concentrations and meteorological fields with chemical reaction rate constants obtained in lab experiments. They are able to identify not only the key chemical reactions but also the transport processes which bring CFC photolysis products into contact with ozone.

The Ozone Hole and its Causes

The Antarctic ozone hole is an area of the Antarctic stratosphere in which the recent ozone levels have dropped to as low as 33% of their pre-1975 values.

The ozone hole occurs during the Antarctic spring, from September to early December, as strong westerly winds start to circulate around the continent and create an atmospheric container. Within this polar vortex, over 50% of the lower stratospheric ozone is destroyed during the Antarctic spring.

As explained above, the primary cause of ozone depletion is the presence of chlorine-containing source gases (primarily CFCs and related halocarbons).

In the presence of UV light, these gases dissociate, releasing chlorine atoms, which then go on to catalyze ozone destruction. The Cl-catalyzed ozone depletion can take place in the gas phase, but it is dramatically enhanced in the presence of polar stratospheric clouds (PSCs).

These polar stratospheric clouds(PSC) form during winter, in the extreme cold. Polar winters are dark, consisting of 3 months without solar radiation (sunlight).

The lack of sunlight contributes to a decrease in temperature and the polar vortex traps and chills air. Temperatures hover around or below -80 °C. These low temperatures form cloud particles. There are three types of PSC clouds; nitric acid trihydrate clouds, slowly cooling water-ice clouds, and rapid cooling water-ice(nacerous) clouds; that provide surfaces for chemical reactions that lead to ozone destruction.

The photochemical processes involved are complex but well understood. The key observation is that, ordinarily, most of the chlorine in the stratosphere resides in stable "reservoir" compounds, primarily hydrochloric acid (HCl) and chlorine nitrate ($ClONO_2$). During the Antarctic winter and spring, however, reactions on the surface of the polar stratospheric cloud particles convert these "reservoir" compounds into reactive free radicals (Cl and ClO). The clouds can also remove NO_2 from the atmosphere by converting it to nitric acid, which prevents the newly formed ClO from being converted back into $ClONO_2$.

The role of sunlight in ozone depletion is the reason why the Antarctic ozone depletion is greatest during spring. During winter, even though PSCs are at their most abundant, there is no light over the pole to drive the chemical reactions. During the spring, however, the sun comes out, providing energy to drive photochemical reactions, and melt the polar stratospheric clouds, releasing the trapped compounds. Warming temperatures near the end of spring break up the vortex around mid-December. As warm, ozone-rich air flows in from lower latitudes, the PSCs are destroyed, the ozone depletion process shuts down, and the ozone hole closes.

Most of the ozone that is destroyed is in the lower stratosphere, in contrast to the much smaller ozone depletion through homogeneous gas phase reactions, which occurs primarily in the upper stratosphere.

Interest in Ozone Layer Depletion

While the effect of the Antarctic ozone hole in decreasing the global ozone is relatively small, estimated at about 4% per decade, the hole has generated a great deal of interest because:

- The decrease in the ozone layer was predicted in the early 1980s to be roughly 7% over a 60 year period.
- The sudden recognition in 1985 that there was a substantial "hole" was widely reported in the press. The especially rapid ozone depletion in Antarctica had previously been dismissed as a measurement error.
- Many were worried that ozone holes might start to appear over other areas of the globe but to date the only other large-scale depletion is a smaller ozone "dimple" observed during the Arctic spring over the North Pole. Ozone at middle latitudes has declined, but by a much smaller extent (about 4–5% decrease).
- If the conditions became more severe (cooler stratospheric temperatures, more stratospheric clouds, more active chlorine), then global ozone may decrease at a much greater pace. Standard global warming theory predicts that the stratosphere will cool.
- When the Antarctic ozone hole breaks up, the ozone-depleted air drifts out into nearby areas. Decreases in the ozone level of up to 10% have been reported in New Zealand in the month following the break-up of the Antarctic ozone hole.

Consequences of Ozone Layer Depletion

Since the ozone layer absorbs UVB ultraviolet light from the Sun, ozone layer depletion is expected to increase surface UVB levels, which could lead to damage, including increases in skin cancer.

This was the reason for the Montreal Protocol. Although decreases in stratospheric ozone are well-tied to CFCs and there are good theoretical reasons to believe that decreases in ozone will lead to increases in surface UVB, there is no direct observational evidence linking ozone depletion to higher incidence of skin cancer in human beings. This is partly because UVA, which has also been implicated in some forms of skin cancer, is not absorbed by ozone, and it is nearly impossible to control statistics for lifestyle changes in the populace.

Increased UV

Ozone, while a minority constituent in the Earth's atmosphere, is responsible for most of the absorption of UVB radiation. The amount of UVB radiation that penetrates through the ozone layer decreases exponentially with the slant-path thickness/density of the layer. Correspondingly, a decrease in atmospheric ozone is expected to give rise to significantly increased levels of UVB near the surface.

Increases in surface UVB due to the ozone hole can be partially inferred by radiative transfer model calculations, but cannot be calculated from direct measurements because of the lack of reliable historical (pre-ozone-hole) surface UV data, although more recent surface UV observation measurement programmes exist (e.g. at Lauder, New Zealand).

Because it is this same UV radiation that creates ozone in the ozone layer from O_2 (regular oxygen) in the first place, a reduction in stratospheric ozone would actually tend to increase photochemical production of ozone at lower levels (in the troposphere), although the overall observed trends in total column ozone still show a decrease, largely because ozone produced lower down has a naturally shorter photochemical lifetime, so it is destroyed before the concentrations could reach a level which would compensate for the ozone reduction higher up.

Biological Effects

The main public concern regarding the ozone hole has been the effects of increased surface UV radiation on human health. So far, ozone depletion in most locations has been typically a few percent and, as noted above, no direct evidence of health damage is available in most latitudes.

Were the high levels of depletion seen in the ozone hole ever to be common across the globe, the effects could be substantially more dramatic. As the ozone hole over Antarctica has in some instances grown so large as to reach southern parts of Australia, New Zealand, Chile, Argentina, and South Africa, environmentalists have been concerned that the increase in surface UV could be significant.

Ozone depletion will change all of the effects of UVB on human health, both positive and negative.

UVB (the higher energy UV radiation absorbed by ozone) is generally accepted to be a contributory factor to skin cancer and to produce Vitamin D. In addition, increased surface UV leads to increased tropospheric ozone, which is a health risk to humans.

1. *Basal and Squamous Cell Carcinomas:* The most common forms of skin cancer in humans, basal and squamous cell carcinomas, have been strongly linked to UVB exposure. The mechanism by which UVB induces these cancers is well understood—absorption of UVB radiation causes the pyrimidine bases in the DNA molecule to form dimers, resulting in transcription errors when the DNA replicates. These cancers are relatively mild and rarely fatal, although the treatment of squamous cell carcinoma sometimes requires extensive reconstructive surgery. By combining epidemiological data with results of animal studies, scientists have estimated that a one percent decrease in stratospheric ozone would increase the incidence of these cancers by 2%.
2. *Malignant Melanoma:* Another form of skin cancer, malignant melanoma, is much less common but far more dangerous, being lethal in about 15–20% of the cases diagnosed.

   The relationship between malignant melanoma and ultraviolet exposure is not yet well understood, but it appears that both UVB and UVA are involved.

   Experiments on fish suggest that 90 to 95% of malignant melanomas may be due to UVA and visible radiation whereas experiments on opossums suggest a larger role for UVB.

   Because of this uncertainty, it is difficult to estimate the impact of ozone depletion on melanoma incidence. One study showed that a 10% increase in UVB radiation was associated with a 19% increase in melanomas for men and 16% for women.

   A study of people in Punta Arenas, at the southern tip of Chile, showed a 56% increase in melanoma and a 46% increase in nonmelanoma skin cancer over a period of seven years, along with decreased ozone and increased UVB levels.
3. *Cortical Cataracts:* Studies are suggestive of an association between ocular cortical cataracts and UV-B exposure, using crude approximations of exposure and various cataract assessment techniques.

   A detailed assessment of ocular exposure to UV-B was carried out in a study on Chesapeake Bay Watermen, where increases in average annual ocular exposure were associated with increasing risk of cortical opacity. In this highly exposed group of predominantly white males, the evidence linking cortical opacities to sunlight exposure was the strongest to date.

However, subsequent data from a population-based study in Beaver Dam, WI suggested the risk may be confined to men. In the Beaver Dam study, the exposures among women were lower than exposures among men, and no association was seen. Moreover, there were no data linking sunlight exposure to risk of cataract in African Americans, although other eye diseases have different prevalences among the different racial groups, and cortical opacity appears to be higher in African Americans compared with whites.

4. *Increased Tropospheric Ozone:* Increased surface UV leads to increased tropospheric ozone.

   Ground-level ozone is generally recognised to be a health risk, as ozone is toxic due to its strong oxidant properties. At this time, ozone at ground level is produced mainly by the action of UV radiation on combustion gases from vehicle exhausts.

5. *Increased production of Vitamin D:* Vitamin D is produced in the skin by ultraviolet light. Thus, higher UV-B exposure raises human vitamin D in those deficient in it.

   Recent research (primarily since the Montreal protocol), shows that many humans have less than optimal vitamin D levels.

   In particular, the lowest quartile of vitamin D (<17.8 ng/ml), in the US population were found using information from the National Health and Nutrition Examination Survey to be associated with an increase in all cause mortality in the general population.

   While, higher level of Vitamin D are associated with higher mortality, the body has mechanisms that prevent sunlight from producing too much Vitamin D.

Effects on Non-Human Animals

A November 2010 report by scientists at the Institute of Zoology in London found that whales off the coast of California have shown a sharp rise in sun damage, and these scientists "fear that the thinning ozone layer is to blame."

The study photographed and took skin biopsies from over 150 whales in the Gulf of California and found "widespread evidence of epidermal damage commonly associated with acute and severe sunburn," having cells which form when the DNA is damaged by UV radiation.

The findings suggest "rising UV levels as a result of ozone depletion are to blame for the observed skin damage, in the same way that human skin cancer rates have been on the increase in recent decades."

Effects on Crops

An increase of UV radiation would be expected to affect crops. A number of economically important species of plants, such as rice, depend on cyanobacteria residing on their roots for the retention of nitrogen. Cyanobacteria are sensitive to UV light and would be affected by its increase.

Public Policy

The full extent of the damage that CFCs have caused to the ozone layer is not known and will not be known for decades; however, marked decreases in column ozone have already been observed (as explained before).

After a 1976 report by the United States National Academy of Sciences concluded that credible scientific evidence supported the ozone depletion hypothesis a few countries, including the United States, Canada, Sweden, Denmark, and Norway, moved to eliminate the use of CFCs in aerosol spray cans.

At the time this was widely regarded as a first step towards a more comprehensive regulation policy, but progress in this direction slowed in subsequent years, due to a combination of political factors (continued resistance from the halocarbon industry and a general change in attitude towards environmental regulation during the first two years of the Reagan administration) and scientific developments (subsequent National Academy assessments which indicated that the first estimates of the magnitude of ozone depletion had been overly large).

A critical DuPont manufacturing patent for Freon was set to expire in 1979 Chlorofluorocarbon#Regulation_and_DuPont. The United States banned the use of CFCs in aerosol cans in 1978. The European Community rejected proposals to ban CFCs in aerosol sprays, and in the U.S., CFCs continued to be used as refrigerants and for cleaning circuit boards. Worldwide CFC production fell sharply after the U.S. aerosol ban, but by 1986 had returned nearly to its 1976 level. In 1993, DuPont shut down its CFC facility.

The U.S. Government's attitude began to change again in 1983, when William Ruckelshaus replaced Anne M. Burford as Administrator of the United States Environmental Protection Agency. Under

Ruckelshaus and his successor, Lee Thomas, the EPA pushed for an international approach to halocarbon regulations. In 1985 20 nations, including most of the major CFC producers, signed the Vienna Convention for the Protection of the Ozone Layer which established a framework for negotiating international regulations on ozone-depleting substances. That same year, the discovery of the Antarctic ozone hole was announced, causing a revival in public attention to the issue. In 1987, representatives from 43 nations signed the Montreal Protocol. Meanwhile, the halocarbon industry shifted its position and started supporting a protocol to limit CFC production.

The reasons for this were in part explained by "Dr. Mostafa Tolba, former head of the UN Environment Programme, who was quoted in the 30 June 1990 edition of The New Scientist, '...the chemical industry supported the Montreal Protocol in 1987 because it set up a worldwide schedule for phasing out CFCs, which [were] no longer protected by patents. This provided companies with an equal opportunity to market new, more profitable compounds.'"

At Montreal, the participants agreed to freeze production of CFCs at 1986 levels and to reduce production by 50% by 1999. After a series of scientific expeditions to the Antarctic produced convincing evidence that the ozone hole was indeed caused by chlorine and bromine from manmade organohalogens, the Montreal Protocol was strengthened at a 1990 meeting in London.

The participants agreed to phase out CFCs and halons entirely (aside from a very small amount marked for certain "essential" uses, such as asthma inhalers) by 2000 in non-Article 5 countries and by 2010 in Article 5 (less developed) signatories. At a 1992 meeting in Copenhagen, the phase out date was moved up to 1996. At the same meeting, methyl bromide (MeBr), a fumigant used primarily in agricultural production, was added to the list of controlled substances. It should be noted that for all substances controlled under the Protocol, phaseout schedules were delayed for less developed ('Article 5(1)') countries, and phaseout in these countries was supported by transfers of expertise, technology, and money from non-Article 5(1) Parties to the Protocol. Additionally, exemptions from the agreed schedules could be applied for under the Essential Use Exemption (EUE) process for substances other than methyl bromide and under the Critical Use Exemption (CUE) process for methyl bromide.

To some extent, CFCs have been replaced by the less damaging hydro-chloro-fluoro-carbons (HCFCs), although concerns remain regarding HCFCs also. In some applications, hydro-fluoro-carbons (HFCs) have been used to replace CFCs. HFCs, which contain no chlorine or bromine, do not contribute at all to ozone depletion although they are potent greenhouse gases. The best known of these compounds is probably HFC-134a (R-134a), which in the United States has largely replaced CFC-12 (R-12) in automobile air conditioners. In laboratory analytics (a former "essential" use) the ozone depleting substances can be replaced with various other solvents.

Ozone Diplomacy, by Richard Benedick (Harvard University Press, 1991) gives a detailed account of the negotiation process that led to the Montreal Protocol. Pielke and Betsill provide an extensive review of early U.S. government responses to the emerging science of ozone depletion by CFCs.

More recently, policy experts have advocated for efforts to link ozone protection efforts to climate protection efforts. Many ODS are also greenhouse gasses, some significantly more powerful agents of radiative forcing than carbon dioxide over the short and medium term. Policy decisions in one arena affect the costs and effectiveness of environmental improvements in the other.

Prospects of Ozone Depletion

Since the adoption and strengthening of the Montreal Protocol has led to reductions in the emissions of CFCs, atmospheric concentrations of the most significant compounds have been declining. These substances are being gradually removed from the atmosphere—since peaking in 1994, the Effective Equivalent Chlorine (EECl) level in the atmosphere had dropped about 10% by 2008. It is estimated that by 2015, the Antarctic ozone hole will have reduced by 1 million km^2 out of 25 (Newman *et al.*, 2004); complete recovery of the Antarctic ozone layer is not expected to occur until the year 2050 or later. Work has suggested that a detectable (and statistically significant) recovery will not occur until around 2024, with ozone levels recovering to 1980 levels by around 2068. The decrease in ozone-depleting chemicals has also been significantly affected by a decrease in bromine-containing chemicals. The data suggest that substantial natural sources exist for atmospheric methyl bromide (CH_3Br). The phase-out of CFCs means that nitrous oxide (N_2O), which is not covered by the Montreal Protocol, has become the most highly emitted ozone depleting substance and is expected to remain so throughout the 21st century.

When the 2004 ozone hole ended in November 2004, daily minimum stratospheric temperatures in the Antarctic lower stratosphere increased to levels that are too warm for the formation of polar stratospheric clouds (PSCs) about 2 to 3 weeks earlier than in most recent years.

The Arctic winter of 2005 was extremely cold in the stratosphere; PSCs were abundant over many high-latitude areas until dissipated by a big warming event, which started in the upper stratosphere during February and spread throughout the Arctic stratosphere in March. The size of the Arctic area of anomalously low total ozone in 2004–2005 was larger than in any year since 1997. The predominance of anomalously low total ozone values in the Arctic region in the winter of 2004–2005 is attributed to the very low stratospheric temperatures and meteorological conditions favourable for ozone destruction along with the continued presence of ozone destroying chemicals in the stratosphere.

A 2005 IPCC summary of ozone issues concluded that observations and model calculations suggest that the global average amount of ozone depletion has now approximately stabilized. Although considerable variability in ozone is expected from year to year, including in polar regions where depletion is largest, the ozone layer is expected to begin to recover in coming decades due to declining ozone-depleting substance concentrations, assuming full compliance with the Montreal Protocol.

Temperatures during the Arctic winter of 2006 stayed fairly close to the long-term average until late January, with minimum readings frequently cold enough to produce PSCs. During the last week of January, however, a major warming event sent temperatures well above normal — much too warm to support PSCs. By the time temperatures dropped back to near normal in March, the seasonal norm was well above the PSC threshold. Preliminary satellite instrument-generated ozone maps show seasonal ozone buildup slightly below the long-term means for the Northern Hemisphere as a whole, although some high ozone events have occurred. During March 2006, the Arctic stratosphere poleward of 60° North Latitude was free of anomalously low ozone areas except during the three-day period from 17 March to 19 when the total ozone cover fell below 300 DU over part of the North Atlantic region from Greenland to Scandinavia.

The area where total column ozone is less than 220 DU (the accepted definition of the boundary of the ozone hole) was relatively small until around 20 August 2006. Since then the ozone hole area increased rapidly, peaking at 29 million km^2 24 September. In October

2006, NASA reported that the year's ozone hole set a new area record with a daily average of 26 million km^2 between 7 September and 13 October 2006; total ozone thicknesses fell as low as 85 DU on 8 October. The two factors combined, 2006 sees the worst level of depletion in recorded ozone history. The depletion is attributed to the temperatures above the Antarctic reaching the lowest recording since comprehensive records began in 1979.

On October 2008 the Ecuadorian Space Agency published a report called HIPERION, a study of the last 28 years data from 10 satellites and dozens of ground instruments around the world among them their own, and found that the UV radiation reaching equatorial latitudes was far greater than expected, climbing in some very populated cities up to 24 UVI, the WHO UV Index standard considers 11 as an extreme index and a great risk to health. The report concluded that the ozone depletion around mid latitudes on the planet is already endangering large populations in this areas. Later, the CONIDA, the Peruvian Space Agency, made its own study, which found almost the same facts as the Ecuadorian study.

The Antarctic ozone hole is expected to continue for decades. Ozone concentrations in the lower stratosphere over Antarctica will increase by 5%–10% by 2020 and return to pre-1980 levels by about 2060–2075, 10–25 years later than predicted in earlier assessments. This is because of revised estimates of atmospheric concentrations of Ozone Depleting Substances — and a larger predicted future usage in developing countries. Another factor which may aggravate ozone depletion is the draw-down of nitrogen oxides from above the stratosphere due to changing wind patterns.

History of the Research

The basic physical and chemical processes that lead to the formation of an ozone layer in the Earth's stratosphere were discovered by Sydney Chapman in 1930. These are discussed in the article Ozone-oxygen cycle — briefly, short-wavelength UV radiation splits an oxygen (O_2) molecule into two oxygen (O) atoms, which then combine with other oxygen molecules to form ozone. Ozone is removed when an oxygen atom and an ozone molecule "recombine" to form two oxygen molecules, i.e. $O + O_3 \rightarrow 2O_2$. In the 1950s, David Bates and Marcel Nicolet presented evidence that various free radicals, in particular hydroxyl (OH) and nitric oxide (NO), could catalyze this recombination reaction, reducing the overall amount of ozone. These free radicals were known to be present in the stratosphere, and so were regarded as part of the natural balance – it was estimated that in their absence, the ozone layer would be about twice as thick as it currently is.

In 1970 Prof. Paul Crutzen pointed out that emissions of *nitrous* oxide (N_2O), a stable, long-lived gas produced by soil bacteria, from the Earth's surface could affect the amount of *nitric* oxide (NO) in the stratosphere.

Crutzen showed that nitrous oxide lives long enough to reach the stratosphere, where it is converted into NO. Crutzen then noted that increasing use of fertilizers might have led to an increase in nitrous oxide emissions over the natural background, which would in turn result in an increase in the amount of NO in the stratosphere.

Thus human activity could have an impact on the stratospheric ozone layer. In the following year, Crutzen and (independently) Harold Johnston suggested that NO emissions from supersonic aircraft, which fly in the lower stratosphere, could also deplete the ozone layer.

The Rowland-Molina Hypothesis

In 1974 Frank Sherwood Rowland, Chemistry Professor at the University of California at Irvine, and his postdoctoral associate Mario J. Molina suggested that long-lived organic halogen compounds, such as CFCs, might behave in a similar fashion as Crutzen had proposed for nitrous oxide. James Lovelock (most popularly known as the creator of the Gaia hypothesis) had recently discovered, during a cruise in the South Atlantic in 1971, that almost all of the CFC compounds manufactured since their invention in 1930 were still present in the atmosphere. Molina and Rowland concluded that, like N_2O, the CFCs would reach the stratosphere where they would be dissociated by UV light, releasing Cl atoms. (A year earlier, Richard Stolarski and Ralph Cicerone at the University of Michigan had shown that Cl is even more efficient than NO at catalyzing the destruction of ozone. Similar conclusions were reached by Michael McElroy and Steven Wofsy at Harvard University.

Neither group, however, had realised that CFCs were a potentially large source of stratospheric chlorine — instead, they had been investigating the possible effects of HCl emissions from the Space Shuttle, which are very much smaller.)

The Rowland-Molina hypothesis was strongly disputed by representatives of the aerosol and halocarbon industries. The Chair of the Board of DuPont was quoted as saying that ozone depletion theory is "a science fiction tale...a load of rubbish...utter nonsense".

Robert Abplanalp, the President of Precision Valve Corporation (and inventor of the first practical aerosol spray can valve), wrote to the Chancellor of UC Irvine to complain about Rowland's public statements (Roan, p 56.) Nevertheless, within three years most of the basic assumptions made by Rowland and Molina were confirmed by laboratory measurements and by direct observation in the stratosphere.

The concentrations of the source gases (CFCs and related compounds) and the chlorine reservoir species (HCl and $ClONO_2$) were measured throughout the stratosphere, and demonstrated that CFCs were indeed the major source of stratospheric chlorine, and that nearly all of the CFCs emitted would eventually reach the stratosphere.

Even more convincing was the measurement, by James G. Anderson and collaborators, of chlorine monoxide (ClO) in the stratosphere. ClO is produced by the reaction of Cl with ozone — its observation thus demonstrated that Cl radicals not only were present in the stratosphere but also were actually involved in destroying ozone.

McElroy and Wofsy extended the work of Rowland and Molina by showing that bromine atoms were even more effective catalysts for ozone loss than chlorine atoms and argued that the brominated organic compounds known as halons, widely used in fire extinguishers, were a potentially large source of stratospheric bromine. In 1976 the United States National Academy of Sciences released a report which concluded that the ozone depletion hypothesis was strongly supported by the scientific evidence. Scientists calculated that if CFC production continued to increase at the going rate of 10% per year until 1990 and then remain steady, CFCs would cause a global ozone loss of 5 to 7% by 1995, and a 30 to 50% loss by 2050. In response the United States, Canada and Norway banned the use of CFCs in aerosol spray cans in 1978. However, subsequent research, summarised by the National Academy in reports issued between 1979 and 1984, appeared to show that the earlier estimates of global ozone loss had been too large.

Crutzen, Molina, and Rowland were awarded the 1995 Nobel Prize in Chemistry for their work on stratospheric ozone.

The Ozone Hole

The discovery of the Antarctic "ozone hole" by British Antarctic Survey scientists Farman, Gardiner and Shanklin (announced in a paper in *Nature* in May 1985) came as a shock to the scientific community, because the observed decline in polar ozone was far larger than anyone had anticipated.

Satellite measurements showing massive depletion of ozone around the south pole were becoming available at the same time. However, these were initially rejected as unreasonable by data quality control algorithms (they were filtered out as errors since the values were unexpectedly low); the ozone hole was detected only in satellite data when the raw data was reprocessed following evidence of ozone depletion in *in situ* observations. When the software was rerun without the flags, the ozone hole was seen as far back as 1976.

Susan Solomon, an atmospheric chemist at the National Oceanic and Atmospheric Administration (NOAA), proposed that chemical reactions on polar stratospheric clouds (PSCs) in the cold Antarctic stratosphere caused a massive, though localized and seasonal, increase in the amount of chlorine present in active, ozone-destroying forms.

The polar stratospheric clouds in Antarctica are only formed when there are very low temperatures, as low as -80 °C, and early spring conditions. In such conditions the ice crystals of the cloud provide a suitable surface for conversion of unreactive chlorine compounds into reactive chlorine compounds which can deplete ozone easily.

Moreover the polar vortex formed over Antarctica is very tight and the reaction which occurs on the surface of the cloud crystals is far different from when it occurs in atmosphere. These conditions have led to ozone hole formation in Antarctica. This hypothesis was decisively confirmed, first by laboratory measurements and subsequently by direct measurements, from the ground and from high-altitude airplanes, of very high concentrations of chlorine monoxide (ClO) in the Antarctic stratosphere.

Alternative hypotheses, which had attributed the ozone hole to variations in solar UV radiation or to changes in atmospheric circulation patterns, were also tested and shown to be untenable.

Meanwhile, analysis of ozone measurements from the worldwide network of ground-based Dobson spectrophotometres led an international panel to conclude that the ozone layer was in fact being depleted, at all latitudes outside of the tropics. These trends were confirmed by satellite measurements. As a consequence, the major halocarbon producing nations agreed to phase out production of CFCs, halons, and related compounds, a process that was completed in 1996.

Since 1981 the United Nations Environment Programme, under the auspices of the World Meteorological Organisation, has sponsored a series of technical reports on the Scientific Assessment of Ozone

Depletion, based on satellite measurements. The 2007 report showed that the hole in the ozone layer was recovering and the smallest it had been for about a decade. The 2010 report found that "Over the past decade, global ozone and ozone in the Arctic and Antarctic regions is no longer decreasing but is not yet increasing... the ozone layer outside the Polar regions is projected to recover to its pre-1980 levels some time before the middle of this century... In contrast, the springtime ozone hole over the Antarctic is expected to recover much later."

Arctic Ozone Hole

On March 15, 2011, a record ozone layer loss was observed, with about half of the ozone present over the Arctic having been destroyed. The change was attributed to increasingly cold winters in the Arctic stratosphere at an altitude of approximately 20 km (12.4 miles), a change associated with global warming in a relationship that is still under investigation. By March 25, the ozone loss had become the largest compared to that observed in all previous winters with the possibility that it would become an ozone hole. This would require that the quantities of ozone to fall below 200 Dobson units, from the 250 recorded over central Siberia. It is predicted that the thinning layer would affect parts of Scandinavia and Eastern Europe on March 30–31.

Tibet Ozone Hole

As winters that are colder are more affected, at times there is an ozone hole over Tibet. In 2006, a 2.5 million square kilometre ozone hole was detected over Tibet. Also again in 2011 an ozone hole appeared over mountainous regions of Tibet, Xinjiang, Qinghai and the Hindu Kush, along with an unprecedented hole over the Arctic, though the Tibet one is far less intense than the ones over the arctic or antarctic.

Ozone Depletion and Global Warming

There are five areas of linkage between ozone depletion and global warming:

- The same CO_2 radiative forcing that produces global warming is expected to cool the stratosphere. This cooling, in turn, is expected to produce a relative *increase* in ozone (O_3) depletion in polar area and the frequency of ozone holes.
- Conversely, ozone depletion represents a radiative forcing of the climate system. There are two opposing effects: Reduced

ozone causes the stratosphere to absorb less solar radiation, thus cooling the stratosphere while warming the troposphere; the resulting colder stratosphere emits less long-wave radiation downward, thus cooling the troposphere. Overall, the cooling dominates; the IPCC concludes that "*observed stratospheric O_3 losses over the past two decades have caused a negative forcing of the surface-troposphere system*" of about –0.15 ± 0.10 watts per square metre (W/m^2).

- One of the strongest predictions of the greenhouse effect is that the stratosphere will cool. Although this cooling has been observed, it is not trivial to separate the effects of changes in the concentration of greenhouse gases and ozone depletion since both will lead to cooling. However, this can be done by numerical stratospheric modelling. Results from the National Oceanic and Atmospheric Administration's Geophysical Fluid Dynamics Laboratory show that above 20 km (12.4 miles), the greenhouse gases dominate the cooling.
- As noted under 'Public Policy', ozone depleting chemicals are also often greenhouse gases. The increases in concentrations of these chemicals have produced 0.34 ± 0.03 W/m^2 of radiative forcing, corresponding to about 14% of the total radiative forcing from increases in the concentrations of well-mixed greenhouse gases.
- The long term modelling of the process, its measurement, study, design of theories and testing take decades to document, gain wide acceptance, and ultimately become the dominant paradigm. Several theories about the destruction of ozone were hypothesized in the 1980s, published in the late 1990s, and are currently being proven. Dr Drew Schindell, and Dr Paul Newman, NASA Goddard, proposed a theory in the late 1990s, using a SGI Origin 2000 supercomputer, that modelled ozone destruction, accounted for 78% of the ozone destroyed. Further refinement of that model accounted for 89% of the ozone destroyed, but pushed back the estimated recovery of the ozone hole from 75 years to 150 years. (An important part of that model is the lack of stratospheric flight due to depletion of fossil fuels.)

Misconceptions about Ozone Depletion

CFCs are "too heavy" to reach the stratosphere: It is commonly believed that CFC molecules are heavier than air (nitrogen or oxygen), so that the CFC molecules cannot reach the stratosphere in significant amount. But atmospheric gases are not sorted by weight; the forces of wind can fully mix the gases in the atmosphere. Despite the fact that CFCs are heavier than air and with a long lifetime, they are evenly distributed throughout the turbosphere and reach the upper atmosphere.

Very violent volcanic eruptions can inject HCl into the stratosphere, but researchers have shown that the contribution is not significant compared to that from CFCs. A similar erroneous assertion is that soluble halogen compounds from the volcanic plume of Mount Erebus on Ross Island, Antarctica are a major contributor to the Antarctic ozone hole.

An ozone hole was first observed in 1956: G.M.B. Dobson (Exploring the Atmosphere, 2nd Edition, Oxford, 1968) mentioned that when springtime ozone levels over Halley Bay were first measured in 1956, he was surprised to find that they were ~320 DU, about 150 DU below spring levels, ~450 DU, in the Arctic.

These, however, were at this time the known normal climatological values because no other Antarctic ozone data were available. What Dobson describes is essentially the *baseline* from which the ozone hole is measured: actual ozone hole values are in the 150–100 DU range.

The discrepancy between the Arctic and Antarctic noted by Dobson was primarily a matter of timing: during the Arctic spring ozone levels rose smoothly, peaking in April, whereas in the Antarctic they stayed approximately constant during early spring, rising abruptly in November when the polar vortex broke down.

The behaviour seen in the Antarctic ozone hole is completely different. Instead of staying constant, early springtime ozone levels suddenly drop from their already low winter values, by as much as 50%, and normal values are not reached again until December.

The ozone hole should be above the sources of CFCs: Some people thought that the ozone hole should be above the sources of CFCs. However, CFCs are well mixed globally in the troposphere and the stratosphere. The reason for occurrence of the ozone hole above Antarctica is not because there are more CFCs concentrated but because the low temperatures help form polar stratospheric clouds. In fact, there are findings of significant and localized "ozone holes" above other parts of the earth.

The "ozone hole" is a hole in the ozone layer: There is a common misconception that the "ozone hole" is really a hole in the ozone layer. When the "ozone hole" occurs, the ozone in the lower stratosphere is destroyed. The upper stratosphere is less affected, so that the amount of ozone over the continent decreases by 50 percent or even more. The ozone does not disappear through the layer, nor is there a uniform 'thinning' of the ozone layer. It is a "hole" which is a depression, not in the sense of "a hole in the windshield."

ODS Requirements in the Marine Industry

IMO has amended MARPOL Annex VI Regulation 12 regarding ozone depleting substances.

As from July 1, 2010, all vessels where MARPOL Annex VI is applicable should have a list of equipment using ozone depleting substances. The list should include name of ODS, type and location of equipment, quantity in kg and date. All changes since that date should be recorded in a ODS Record book on board recording all intended or unintended releases to the atmosphere. Furthermore, new ODS supply or landing to shore facilities should be recorded as well.

World Ozone Day

In 1994, the United Nations General Assembly voted to designate September 16 as "World Ozone Day", to commemorate the signing of the Montreal Protocol on that date in 1987.

4

Climatology

Climatology (from Greek, *klima*, "place, zone"; and *-logia*) is the study of climate, scientifically defined as weather conditions averaged over a period of time, and is a branch of the atmospheric sciences. Basic knowledge of climate can be used within shorter term weather forecasting using analog techniques such as the El Niño – Southern Oscillation (ENSO), the Madden-Julian Oscillation (MJO), the North Atlantic Oscillation (NAO), the Northern Annualar Mode (NAM), the Arctic oscillation (AO), the Northern Pacific (NP) Index, the Pacific Decadal Oscillation (PDO), and the Interdecadal Pacific Oscillation (IPO). Climate models are used for a variety of purposes from study of the dynamics of the weather and climate system to projections of future climate.

History

The earliest person to hypothesize the concept of climate change may have been the medieval Chinese scientist Shen Kuo (1031–95). Shen Kuo theorized that climates naturally shifted over an enormous span of time, after observing petrified bamboos found underground near Yanzhou (modern day Yan'an, Shaanxi province), a dry climate area unsuitable for the growth of bamboo.

Early climate researchers include Edmund Halley, who published a map of the trade winds in 1686, after a voyage to the southern hemisphere. Benjamin Franklin, in the 18th century, was the first to map the course of the Gulf Stream for use in sending mail overseas from the United States to Europe. Francis Galton invented the term *anticyclone*. Helmut Landsberg led to statistical analysis being used in climatology, which led to its evolution into a physical sciences.

Different Approaches

Climatology is approached in a variety of ways. Paleoclimatology seeks to reconstruct past climates by examining records such as ice cores and tree rings (dendroclimatology). Paleotempestology uses these same records to help determine hurricane frequency over millennia. The study of contemporary climates incorporates meteorological data accumulated over many years, such as records of rainfall, temperature and atmospheric composition. Knowledge of the atmosphere and its dynamics is also embodied in models, either statistical or mathematical, which help by integrating different observations and testing how they fit together. Modelling is used for understanding past, present and potential future climates. Historical climatology is the study of climate as related to human history and thus focuses only on the last few thousand years.

Climate research is made difficult by the large scale, long time periods, and complex processes which govern climate. Climate is governed by physical laws which can be expressed as differential equations. These equations are coupled and nonlinear, so that approximate solutions are obtained by using numerical methods to create global climate models. Climate is sometimes modelled as a stochastic process but this is generally accepted as an approximation to processes that are otherwise too complicated to analyze.

Indices

Scientists use climate indices based on several climate patterns (known as modes of variability) in their attempt to characterise and understand the various climate mechanisms that culminate in our daily weather. Much in the way the Dow Jones Industrial Average, which is based on the stock prices of 30 companies, is used to represent the fluctuations in the stock market as a whole, climate indices are used to represent the essential elements of climate. Climate indices are generally devised with the twin objectives of simplicity and completeness, and each index typically represents the status and timing of the climate factor it represents. By their very nature, indices are simple, and combine many details into a generalised, overall description of the atmosphere or ocean which can be used to characterise the factors which impact the global climate system.

El Niño – Southern Oscillation

El Niño-Southern Oscillation (ENSO) is a global coupled ocean-atmosphere phenomenon. The Pacific ocean signatures, El Niño and

La Niña are important temperature fluctuations in surface waters of the tropical Eastern Pacific Ocean.

The name El Niño, from the Spanish for "the little boy", refers to the Christ child, because the phenomenon is usually noticed around Christmas time in the Pacific Ocean off the west coast of South America.

La Niña means "the little girl". Their effect on climate in the subtropics and the tropics are profound. The atmospheric signature, the Southern Oscillation (SO) reflects the monthly or seasonal fluctuations in the air pressure difference between Tahiti and Darwin. The most recent occurrence of El Niño started in September 2006 and lasted until early 2007.

ENSO is a set of interacting parts of a single global system of coupled ocean-atmosphere climate fluctuations that come about as a consequence of oceanic and atmospheric circulation. ENSO is the most prominent known source of inter-annual variability in weather and climate around the world. The cycle occurs every two to seven years, with El Niño lasting nine months to two years within the longer term cycle, though not all areas globally are affected. ENSO has signatures in the Pacific, Atlantic and Indian Oceans. El Niño causes weather patterns which causes it to rain in specific places but not in others, this is one of many causes for the drought.

In the Pacific, during major warm events, El Niño warming extends over much of the tropical Pacific and becomes clearly linked to the SO intensity. While ENSO events are basically in phase between the Pacific and Indian Oceans, ENSO events in the Atlantic Ocean lag behind those in the Pacific by 12–18 months. Many of the countries most affected by ENSO events are developing countries within tropical sections of continents with economies that are largely dependent upon their agricultural and fishery sectors as a major source of food supply, employment, and foreign exchange. New capabilities to predict the onset of ENSO events in the three oceans can have global socio-economic impacts. While ENSO is a global and natural part of the Earth's climate, whether its intensity or frequency may change as a result of global warming is an important concern. Low-frequency variability has been evidenced: the quasi-decadal oscillation (QDO). Inter-decadal (ID) modulation of ENSO (from PDO or IPO) might exist. This could explain the so-called protracted ENSO of the early 1990s.

North Atlantic Oscillation (NAO)

Indices of the NAO are based on the difference of normalized sea level pressure (SLP) between Ponta Delgada, Azores and

Stykkisholmur/Reykjavik, Iceland. The SLP anomalies at each station were normalized by division of each seasonal mean pressure by the long-term mean (1865–1984) standard deviation. Normalization is done to avoid the series of being dominated by the greater variability of the northern of the two stations. Positive values of the index indicate stronger-than-average westerlies over the middle latitudes.

Northern Pacific (NP) Index

The NP Index is the area-weighted sea level pressure over the region 30N–65N, 160E–140W.

Pacific Decadal Oscillation (PDO)

The PDO is a pattern of Pacific climate variability that shifts phases on at least inter-decadal time scale, usually about 20 to 30 years. The PDO is detected as warm or cool surface waters in the Pacific Ocean, north of 20° N. During a "warm", or "positive", phase, the west Pacific becomes cool and part of the eastern ocean warms; during a "cool" or "negative" phase, the opposite pattern occurs. The mechanism by which the pattern lasts over several years has not been identified; one suggestion is that a thin layer of warm water during summer may shield deeper cold waters. A PDO signal has been reconstructed to 1661 through tree-ring chronologies in the Baja California area.

Composition and Structure of the Atmosphere

An atmosphere (New Latin *atmosphaera*, created in the 17th century from Greek ôìüò [*atmos*] "vapor" and óöáÖñá [*sphaira*] "sphere") is a layer of gases that may surround a material body of sufficient mass, and that is held in place by the gravity of the body. An atmosphere may be retained for a longer duration, if the gravity is high and the atmosphere's temperature is low. Some planets consist mainly of various gases, but only their outer layer is their atmosphere.

The term stellar atmosphere describes the outer region of a star, and typically includes the portion starting from the opaque photosphere outwards. Relatively low-temperature stars may form compound molecules in their outer atmosphere.

Earth's atmosphere, which contains oxygen used by most organisms for respiration and carbon dioxide used by plants, algae and cyanobacteria for photosynthesis, also protects living organisms from genetic damage by solar ultraviolet radiation. Its current composition is the product of billions of years of biochemical modification of the paleoatmosphere by living organisms.

Pressure

Atmospheric pressure is the force of per unit area that is applied perpendicularly to a surface by the surrounding gas. It is determined by a planet's gravitational force in combination with the total mass of a column of air above a location. Units of air pressure are based on the internationally-recognised standard atmosphere (atm), which is defined as 101,325 Pa (or 1,013,250 dynes per cm^2).

The pressure of an atmospheric gas decreases with altitude due to the diminishing mass of gas above each location. The height at which the pressure from an atmosphere declines by a factor of e (an irrational number with a value of 2.71828..) is called the scale height and is denoted by H. For an atmosphere with a uniform temperature, the scale height is proportional to the temperature and inversely proportional to the mean molecular mass of dry air times the planet's gravitational acceleration. For such a model atmosphere, the pressure declines exponentially with increasing altitude. However, atmospheres are not uniform in temperature, so the exact determination of the atmospheric pressure at any particular altitude is more complex.

Escape

Surface gravity, the force that holds down an atmosphere, differs significantly among the planets. For example, the large gravitational force of the giant planet Jupiter is able to retain light gases such as hydrogen and helium that escape from lower gravity objects.

Second, the distance from the sun determines the energy available to heat atmospheric gas to the point where its molecules' thermal motion exceed the planet's escape velocity, the speed at which gas molecules overcome a planet's gravitational grasp. Thus, the distant and cold Titan, Triton, and Pluto are able to retain their atmospheres despite relatively low gravities. Interstellar planets, theoretically, may also retain thick atmospheres.

Since a gas at any particular temperature will have molecules moving at a wide range of velocities, there will almost always be some slow leakage of gas into space. Lighter molecules move faster than heavier ones with the same thermal kinetic energy, and so gases of low molecular weight are lost more rapidly than those of high molecular weight.

It is thought that Venus and Mars may have both lost much of their water when, after being photo dissociated into hydrogen and oxygen by solar ultraviolet, the hydrogen escaped. Earth's magnetic field helps to prevent this, as, normally, the solar wind would greatly enhance the escape of hydrogen. However, over the past 3 billion years the Earth may have lost gases through the magnetic polar regions due to auroral activity, including a net 2% of its atmospheric oxygen.

Other mechanisms that can cause atmosphere depletion are solar wind-induced sputtering, impact erosion, weathering, and sequestration — sometimes referred to as "freezing out" — into the regolith and polar caps.

Composition

Initial atmospheric makeup is generally related to the chemistry and temperature of the local solar nebula during planetary formation and the subsequent escape of interior gases. These original atmospheres underwent much evolution over time, with the varying properties of each planet resulting in very different outcomes.

The atmospheres of the planets Venus and Mars are primarily composed of carbon dioxide, with small quantities of nitrogen, argon, oxygen and traces of other gases.

The atmospheric composition on Earth is largely governed by the by-products of the very life that it sustains. Earth's atmosphere contains roughly (by molar content/volume) 78.08% nitrogen, 20.95% oxygen, a variable amount (average around 1.247%, National Centre for Atmospheric Research) water vapor, 0.93% argon, 0.038% carbon dioxide, and traces of hydrogen, helium, and other "noble" gases.

The low temperatures and higher gravity of the gas giants — Jupiter, Saturn, Uranus and Neptune — allows them to more readily retain gases with low molecular masses. These planets have hydrogen-helium atmospheres, with trace amounts of more complex compounds.

Two satellites of the outer planets possess non-negligible atmospheres: Titan, a moon of Saturn, and Triton, a moon of Neptune, which are mainly nitrogen.

Pluto, in the nearer part of its orbit, has an atmosphere of nitrogen and methane similar to Triton's, but these gases are frozen when farther from the Sun.

Other bodies within the Solar System have extremely thin atmospheres not in equilibrium. These include the Moon (sodium gas), Mercury (sodium gas), Europa (oxygen), Io (sulfur), and Enceladus (water vapor).

The atmospheric composition of an extra-solar planet was first determined using the Hubble Space Telescope. Planet HD 209458b is a gas giant with a close orbit around a star in the constellation Pegasus. The atmosphere is heated to temperatures over 1,000 K, and is steadily escaping into space. Hydrogen, oxygen, carbon and sulfur have been detected in the planet's inflated atmosphere.

Structure

Earth

The Earth's atmosphere consists, from the ground up, of the troposphere (which includes the planetary boundary layer or peplosphere as lowest layer), stratosphere (which includes the ozone layer), mesosphere, thermosphere (which contains the ionosphere), exosphere and also the magnetosphere. Each of the layers has a different lapse rate, defining the rate of change in temperature with height. Three quarters of the atmosphere lies within the troposphere, and the depth of this layer varies between 17 km at the equator and 7 km at the poles. The ozone layer, which absorbs ultraviolet energy from the Sun, is located primarily in the stratosphere, at altitudes of 15 to 35 km. The Kármán line, located within the thermosphere at an altitude of 100 km, is commonly used to define the boundary between the Earth's atmosphere and outer space. However, the exosphere can extend from 500 up to 10,000 km above the surface, where it interacts with the planet's magnetosphere.

The atmosphere of Earth is a layer of gases surrounding the planet Earth that is retained by Earth's gravity. The atmosphere protects life on Earth by absorbing ultraviolet solar radiation, warming the surface through heat retention (greenhouse effect), and reducing temperature extremes between day and night (the diurnal temperature variation). Atmospheric stratification describes the structure of the atmosphere, dividing it into distinct layers, each with specific characteristics such as temperature or composition. The atmosphere has a mass of about 5×10^{18} kg, three quarters of which is within about 11 km (6.8 mi; 36,000 ft) of the surface. The atmosphere becomes thinner and thinner with increasing altitude, with no definite boundary between the atmosphere and outer space. An altitude of 120 km (75 mi) is where atmospheric effects become noticeable during atmospheric reentry of spacecraft. The Kármán line, at 100 km (62 mi), also is often regarded as the boundary between atmosphere and outer space.

Air is the name given to atmosphere used in breathing and photosynthesis. Dry air contains roughly (by volume) 78.09% nitrogen, 20.95% oxygen, 0.93% argon, 0.039% carbon dioxide, and small amounts of other gases. Air also contains a variable amount of water vapor, on average around 1%. While air content and atmospheric pressure varies at different layers, air suitable for the survival of terrestrial plants and terrestrial animals is currently only known to be found in Earth's troposphere and artificial atmospheres.

Composition

Air is mainly composed of nitrogen, oxygen, and argon, which together constitute the major gases of the atmosphere. The remaining gases are often referred to as trace gases, among which are the greenhouse gases such as water vapor, carbon dioxide, methane, nitrous oxide, and ozone. Filtered air includes trace amounts of many other chemical compounds. Many natural substances may be present in tiny amounts in an unfiltered air sample, including dust, pollen and spores, sea spray, and volcanic ash. Various industrial pollutants also may be present, such as chlorine (elementary or in compounds), fluorine compounds, elemental mercury, and sulfur compounds such as sulfur dioxide [SO_2].

Table: *Composition of dry atmosphere, by volume.*

Gas	***Volume***
Nitrogen (N_2)	780,840 ppmv (78.084%)
Oxygen (O_2)	209,460 ppmv (20.946%)
Argon (Ar)	9,340 ppmv (0.9340%)
Carbon dioxide (CO_2)	390 ppmv (0.039%)
Neon (Ne)	18.18 ppmv (0.001818%)
Helium (He)	5.24 ppmv (0.000524%)
Methane (CH_4)	1.79 ppmv (0.000179%)
Krypton (Kr)	1.14 ppmv (0.000114%)
Hydrogen (H_2)	0.55 ppmv (0.000055%)
Nitrous oxide (N_2O)	0.3 ppmv (0.00003%)
Carbon monoxide (CO)	0.1 ppmv (0.00001%)
Xenon (Xe)	0.09 ppmv (9×10^{-6}%) (0.000009%)
Ozone (O_3)	0.0 to 0.07 ppmv (0 to 7×10^{-6}%)
Nitrogen dioxide (NO_2)	0.02 ppmv (2×10^{-6}%) (0.000002%)
Iodine (I_2)	0.01 ppmv (1×10^{-6}%) (0.000001%)
Ammonia (NH_3)	trace
Not included in above dry atmosphere:	
Water vapor (H_2O)	~0.40% over full atmosphere, typically 1%-4% at surface

Structure of the Atmosphere

Principal Layers

In general, air pressure and density decrease in the atmosphere as height increases. However, temperature has a more complicated profile with altitude. Because the general pattern of this profile is constant and recognisable through means such as balloon soundings, temperature provides a useful metric to distinguish between atmospheric layers. In this way, Earth's atmosphere can be divided into five main layers. From highest to lowest, these layers are:

Exosphere

The outermost layer of Earth's atmosphere extends from the exobase upward. It is mainly composed of hydrogen and helium. The particles are so far apart that they can travel hundreds of kilometres without colliding with one another. Since the particles rarely collide, the atmosphere no longer behaves like a fluid. These free-moving particles follow ballistic trajectories and may migrate into and out of the magnetosphere or the solar wind.

Thermosphere

Temperature increases with height in the thermosphere from the mesopause up to the thermopause, then is constant with height. Unlike in the stratosphere, where the inversion is caused by absorption of radiation by ozone, in the thermosphere the inversion is a result of the extremely low density of molecules.

The temperature of this layer can rise to 1,500 °C (2,700 °F), though the gas molecules are so far apart that temperature in the usual sense is not well defined. The air is so rarefied, that an individual molecule (of oxygen, for example) travels an average of 1 kilometre between collisions with other molecules.

The International Space Station orbits in this layer, between 320 and 380 km (200 and 240 mi). Because of the relative infrequency of molecular collisions, air above the mesopause is poorly mixed compared to air below.

While the composition from the troposphere to the mesosphere is fairly constant, above a certain point, air is poorly mixed and becomes compositionally stratified. The point dividing these two regions is known as the turbopause. The region below is the homosphere, and the region above is the heterosphere. The top of the thermosphere is the bottom of the exosphere, called the exobase. Its height varies with solar activity and ranges from about 350–800 km (220–500 mi; 1,100,000–2,600,000 ft).

Mesosphere

The mesosphere extends from the stratopause to 80–85 km (50–53 mi; 260,000–280,000 ft). It is the layer where most meteors burn up upon entering the atmosphere. Temperature decreases with height in the mesosphere.

The mesopause, the temperature minimum that marks the top of the mesosphere, is the coldest place on Earth and has an average temperature around –85 °C (–120 °F; 190 K). At the mesopause, temperatures may drop to –100 °C (–150 °F; 170 K). Due to the cold temperature of the mesosphere, water vapor is frozen, forming ice clouds (or Noctilucent clouds). A type of lightning referred to as either sprites or ELVES, form many miles above thunderclouds in the troposphere.

Stratosphere

The stratosphere extends from the tropopause to about 51 km (32 mi; 170,000 ft). Temperature increases with height due to increased absorption of ultraviolet radiation by the ozone layer, which restricts turbulence and mixing. While the temperature may be –60 °C (–76 °F; 210 K) at the tropopause, the top of the stratosphere is much warmer, and may be near freezing. The stratopause, which is the boundary between the stratosphere and mesosphere, typically is at 50 to 55 km (31 to 34 mi; 160,000 to 180,000 ft). The pressure here is 1/1000 sea level.

Troposphere

The troposphere begins at the surface and extends to between 9 km (30,000 ft) at the poles and 17 km (56,000 ft) at the equator, with some variation due to weather. The troposphere is mostly heated by transfer of energy from the surface, so on average the lowest part of the troposphere is warmest and temperature decreases with altitude. This promotes vertical mixing (hence the origin of its name in the Greek word "ôñïðÞ", *trope*, meaning turn or overturn). The troposphere contains roughly 80% of the mass of the atmosphere. The tropopause is the boundary between the troposphere and stratosphere.

Other Layers

Within the five principal layers determined by temperature are several layers determined by other properties.

- The ozone layer is contained within the stratosphere. In this layer ozone concentrations are about 2 to 8 parts per million, which is much higher than in the lower atmosphere but still very small compared to the main components of the atmosphere.

  It is mainly located in the lower portion of the stratosphere from about 15–35 km (9.3–22 mi; 49,000 –110,000 ft), though the thickness varies seasonally and geographically. About 90% of the ozone in our atmosphere is contained in the stratosphere.
- The ionosphere, the part of the atmosphere that is ionized by solar radiation, stretches from 50 to 1,000 km (31 to 620 mi; 160,000 to 3,300,000 ft) and typically overlaps both the exosphere and the thermosphere.

  It forms the inner edge of the magnetosphere. It has practical importance because it influences, for example, radio propagation on the Earth. It is responsible for auroras.
- The homosphere and heterosphere are defined by whether the atmospheric gases are well mixed.

  In the homosphere the chemical composition of the atmosphere does not depend on molecular weight because the gases are mixed by turbulence.

  The homosphere includes the troposphere, stratosphere, and mesosphere. Above the *turbopause* at about 100 km (62 mi; 330,000 ft) (essentially corresponding to the mesopause), the composition varies with altitude.

  This is because the distance that particles can move without colliding with one another is large compared with the size of motions that cause mixing.

  This allows the gases to stratify by molecular weight, with the heavier ones such as oxygen and nitrogen present only near the bottom of the heterosphere. The upper part of the heterosphere is composed almost completely of hydrogen, the lightest element.
- The planetary boundary layer is the part of the troposphere that is nearest the Earth's surface and is directly affected by it, mainly through turbulent diffusion. During the day the planetary boundary layer usually is well-mixed, while at night it becomes stably stratified with weak or intermittent mixing.

The depth of the planetary boundary layer ranges from as little as about 100 m on clear, calm nights to 3000 m or more during the afternoon in dry regions.

The average temperature of the atmosphere at the surface of Earth is 14 °C (57 °F; 287 K) or 15 °C (59 °F; 288 K), depending on the reference.

Physical Properties

Pressure and Thickness

The average atmospheric pressure at sea level is about 1 atmosphere (atm) = 101.3 kPa (kilopascals) = 14.7 psi (pounds per square inch) = 760 torr = 29.92 inches of mercury (symbol Hg). Total atmospheric mass is 5.1480×10^{18} kg (1.135×10^{19} lb), about 2.5% less than would be inferred from the average sea level pressure and the Earth's area of 51007.2 megahectares, this portion being displaced by the Earth's mountainous terrain.

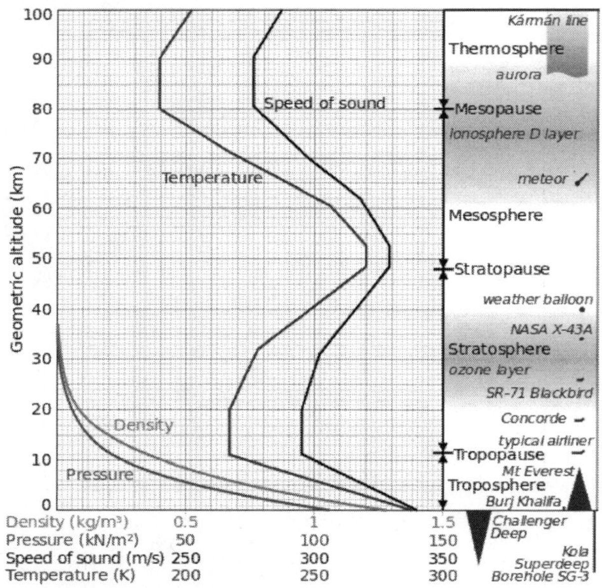

Figure: *Comparison of the 1962 US Standard Atmosphere graph of geometric altitude against air density, pressure, the speed of sound and temperature with approximate altitudes of various objects.*

Atmospheric pressure is the total weight of the air above unit area at the point where the pressure is measured. Thus air pressure varies with location and weather.

If atmospheric density were to remain constant with height the atmosphere would terminate abruptly at 8.50 km (27,900 ft). Instead, density decreases with height, dropping by 50% at an altitude of about 5.6 km (18,000 ft). As a result the pressure decrease is approximately exponential with height, so that pressure decreases by a factor of two approximately every 5.6 km (18,000 ft) and by a factor of $e = 2.718...$ approximately every 7.64 km (25,100 ft), the latter being the average scale height of Earth's atmosphere below 70 km (43 mi; 230,000 ft). However, because of changes in temperature, average molecular weight, and gravity throughout the atmospheric column, the dependence of atmospheric pressure on altitude is modelled by separate equations for each of the layers listed above. Even in the exosphere, the atmosphere is still present. This can be seen by the effects of atmospheric drag on satellites.

In summary, the equations of pressure by altitude in the above references can be used directly to estimate atmospheric thickness. However, the following published data are given for reference:

- 50% of the atmosphere by mass is below an altitude of 5.6 km (18,000 ft).
- 90% of the atmosphere by mass is below an altitude of 16 km (52,000 ft). The common altitude of commercial airliners is about 10 km (33,000 ft) and Mt. Everest's summit is 8,848 m (29,029 ft) above sea level.
- 99.99997% of the atmosphere by mass is below 100 km (62 mi; 330,000 ft), although in the rarefied region above this there are auroras and other atmospheric effects. The highest X-15 plane flight in 1963 reached an altitude of 108.0 km (354,300 ft).

Density and Mass

The density of air at sea level is about 1.2 kg/m^3 (1.2 g/L). Density is not measured directly but is calculated from measurements of temperature, pressure and humidity using the equation of state for air (a form of the ideal gas law). Atmospheric density decreases as the altitude increases. This variation can be approximately modelled using the barometric formula. More sophisticated models are used to predict orbital decay of satellites.

The average mass of the atmosphere is about 5 quadrillion (5×10^{15}) tonnes or 1/1,200,000 the mass of Earth. According to the American National Centre for Atmospheric Research, "The total mean mass of the atmosphere is 5.1480 × 10^{18} kg with an annual range due to water vapor of 1.2 or 1.5 × 10^{15} kg depending on whether surface pressure or water vapor data are used; somewhat smaller than the previous estimate. The mean mass of water vapor is estimated as 1.27×10^{16} kg and the dry air mass as 5.1352 ±0.0003×10^{18} kg."

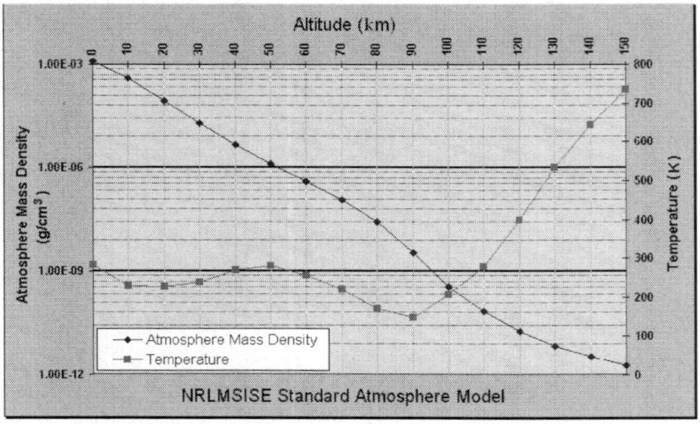

Figure: *Temperature and mass density against altitude from the* ***NRLMSISE-00*** *standard atmosphere model (the eight dotted lines in each "decade" are at the eight cubes 8, 27, 64, ..., 729)*

Optical Properties

Solar radiation (or sunlight) is the energy the Earth receives from the Sun. The Earth also emits radiation back into space, but at longer wavelengths that we cannot see. Part of the incoming and emitted radiation is absorbed or reflected by the atmosphere.

Scattering

When light passes through our atmosphere, photons interact with it through *scattering*. If the light does not interact with the atmosphere, it is called *direct radiation* and is what you see if you were to look directly at the Sun. *Indirect radiation* is light that has been scattered in the atmosphere. For example, on an overcast day when you cannot see your shadow there is no direct radiation reaching you, it has all been scattered. As another example, due to a phenomenon called Rayleigh scattering, shorter (blue) wavelengths scatter more easily than longer (red) wavelengths. This is why the sky looks blue, you are seeing scattered blue light. This is also why sunsets are red. Because the Sun is close to the horizon, the Sun's rays pass through more atmosphere than normal to reach your eye. Much of the blue light has been scattered out, leaving the red light in a sunset.

Absorption

Different molecules absorb different wavelengths of radiation. For example, O_2 and O_3 absorb almost all wavelengths shorter than 300 nanometres. Water (H_2O) absorbs many wavelengths above 700 nm. When a molecule absorbs a photon, it increases the energy of the molecule. We can think of this as heating the atmosphere, but the atmosphere also cools by emitting radiation, as discussed below.

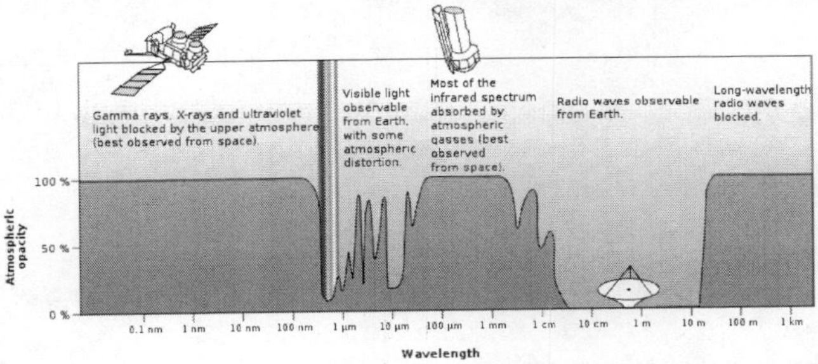

Figure: *Rough plot of Earth's atmospheric transmittance (or opacity) to various wavelengths of electromagnetic radiation, including visible light.*

The combined absorption spectra of the gases in the atmosphere leave "windows" of low opacity, allowing the transmission of only certain bands of light. The optical window runs from around 300 nm (ultraviolet-C) up into the range humans can see, the visible spectrum (commonly called light), at roughly 400–700 nm and continues to the infrared to around 1100 nm. There are also infrared and radio windows that transmit some infrared and radio waves at longer wavelengths. For example, the radio window runs from about one centimetre to about eleven-metre waves.

Emission

Emission is the opposite of absorption, it is when an object emits radiation. Objects tend to emit amounts and wavelengths of radiation depending on their "black body" emission curves, therefore hotter objects tend to emit more radiation, with shorter wavelengths. Colder objects emit less radiation, with longer wavelengths. For example, the Sun is approximately 6,000 K (5,730 °C; 10,340 °F), its radiation peaks near 500 nm, and is visible to the human eye. The Earth is approximately 290 K (17 °C; 62 °F), so its radiation peaks near 10,000 nm, and is much too long to be visible to humans. Because of its temperature, the atmosphere emits infrared radiation. For example, on clear nights the Earth's surface cools down faster than on cloudy nights. This is because clouds (H_2O) are strong absorbers and emitters of infrared radiation. This is also why it becomes colder at night at higher elevations. The atmosphere acts as a "blanket" to limit the amount of radiation the Earth loses into space.

The *greenhouse effect* is directly related to this absorption and emission (or "blanket") effect. Some chemicals in the atmosphere absorb and emit infrared radiation, but do not interact with sunlight in the visible spectrum. Common examples of these chemicals are CO_2 and H_2O. If there are too much of these *greenhouse gases*, sunlight heats the Earth's surface, but the gases block the infrared radiation from exiting back to space. This imbalance causes the Earth to warm, and thus climate change.

Refractive Index

The refractive index of air is close to, but just greater than 1. Systematic variations in refractive index can lead to the bending of light rays over long optical paths. One example is that, under some circumstances, observers onboard ships can see other vessels just over the horizon because light is refracted in the same direction as the curvature of the Earth's surface. The refractive index of air depends on temperature, giving rise to refraction effects when the temperature gradient is large. An example of such effects is the mirage.

Circulation

Atmospheric circulation is the large-scale movement of air through the troposphere, and the means (with ocean circulation) by which heat is distributed around the Earth. The large-scale structure of the atmospheric circulation varies from year to year, but the basic structure remains fairly constant as it is determined by the Earth's rotation rate and the difference in solar radiation between the equator and poles.

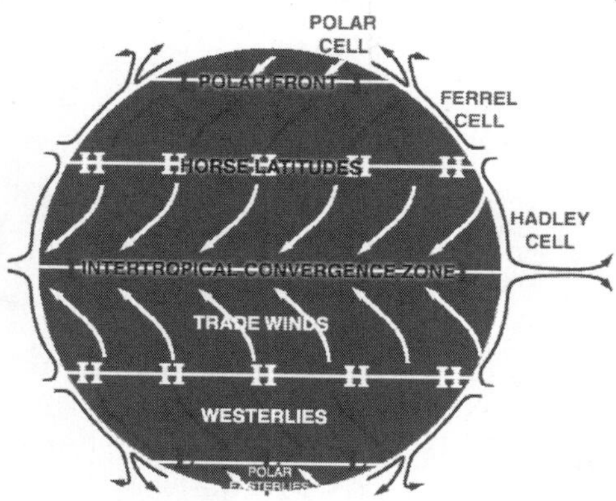

Figure: *An idealised view of three large circulation cells.*

Evolution of Earth's atmosphere

Earliest Atmosphere

The outgassings of the Earth were stripped away by solar winds early in the history of the planet until a steady state was established, the first atmosphere.

Based on today's volcanic evidence, this atmosphere would have contained 60% hydrogen, 20% oxygen (mostly in the form of water vapor), 10% carbon dioxide, 5 to 7% hydrogen sulfide, and smaller amounts of nitrogen, carbon monoxide, free hydrogen, methane and inert gases.

A major rainfall led to the buildup of a vast ocean, enriching the other agents, first carbon dioxide and later nitrogen and inert gases. A major part of carbon dioxide exhalations were soon dissolved in water and built up carbonate sediments.

Second Atmosphere

Water-related sediments have been found dating from as early as 3.8 billion years ago. About 3.4 billion years ago, nitrogen was the major part of the then stable "second atmosphere". An influence of life has to be taken into account rather soon in the history of the atmosphere, since hints of early life forms are to be found as early as 3.5 billion years ago. The fact that this is not perfectly in line with the 30% lower solar radiance (compared to today) of the early Sun has been described as the "faint young Sun paradox".

The geological record however shows a continually relatively warm surface during the complete early temperature record of the Earth with the exception of one cold glacial phase about 2.4 billion years ago. In the late Archaean eon an oxygen-containing atmosphere began to develop, apparently from photosynthesizing algae which have been found as stromatolite fossils from 2.7 billion years ago. The early basic carbon isotopy (isotope ratio proportions) is very much in line with what is found today, suggesting that the fundamental features of the carbon cycle were established as early as 4 billion years ago.

Third Atmosphere

The accretion of continents about 3.5 billion years ago added plate tectonics, constantly rearranging the continents and also shaping long-term climate evolution by allowing the transfer of carbon dioxide to large land-based carbonate storages. Free oxygen did not exist until about 1.7 billion years ago and this can be seen with the development of the red beds and the end of the banded iron formations. This signifies a shift from a reducing atmosphere to an oxidising atmosphere. O_2 showed major ups and downs until reaching a steady state of more than 15%. The following time span was the Phanerozoic eon, during which oxygen-breathing metazoan life forms began to appear. Currently, anthropogenic greenhouse gases are increasing in the atmosphere. According to the Intergovern-mental Panel on Climate Change, this increase is the main cause of global warming.

Air Pollution

Air pollution is the introduction of chemicals, particulate matter, or biological materials that cause harm or discomfort to organisms into the atmosphere. Stratospheric ozone depletion is believed to be caused by air pollution (chiefly from chlorofluorocarbons).

Importance

From the perspective of the planetary geologist, the atmosphere is an evolutionary agent essential to the morphology of a planet. The wind transports dust and other particles which erodes the relief and leaves deposits (eolian processes). Frost and precipitations, which depend on the composition, also influence the relief. Climate changes can influence a planet's geological history. Conversely, studying surface of earth leads to an understanding of the atmosphere and climate of a planet - both its present state and its past.

For a meteorologist, the composition of the atmosphere determines the climate and its variations. For a biologist, the composition is closely dependent on the appearance of the life and its evolution.

Insolation

Insolation is a measure of solar radiation energy received on a given surface area in a given time. It is commonly expressed as average irradiance in watts per square metre (W/m^2) or kilowatt-hours per square metre per day ($kW \cdot h/(m^2 \cdot day)$) (or hours/day). In the case of photovoltaics it is commonly measured as $kWh/(kW_p \cdot y)$ (kilowatt hours per year per kilowatt peak rating). The object or surface that solar radiation strikes may be a planet, a terrestrial object inside the atmosphere of a planet, or any object exposed to solar rays outside of an atmosphere, including spacecraft. Some of the solar radiation will be absorbed, while the remainder will be reflected. Usually the absorbed solar radiation is converted to thermal energy, causing an increasing in the object's temperature. Some systems, however, may store or convert a portion of the solar energy into another form of energy, as in the case of photovoltaics or plants. The proportion of radiation reflected or absorbed depends on the object's reflectivity or albedo.

Projection Effect

The insolation into a surface is largest when the surface directly faces the Sun. As the angle increases between the direction at a right angle to the surface and the direction of the rays of sunlight, the insolation is reduced in proportion to the cosine of the angle.

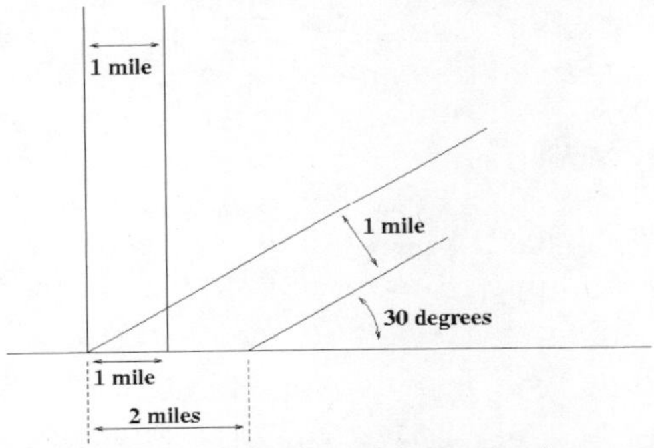

Figure: *One sunbeam one mile wide shines on the ground at a 90° angle, and another at a 30° angle. The oblique sunbeam distributes its light energy over twice as much area.*

In this illustration, the angle shown is between the ground and the sunbeam rather than between the vertical direction and the sunbeam; hence the sine rather than the cosine is appropriate. A sunbeam one mile (1.6 km) wide falls on the ground from directly overhead, and another hits the ground at a 30° angle to the horizontal. Trigonometry tells us that the sine of a 30° angle is 1/2, whereas the sine of a 90° angle is 1. Therefore, the sunbeam hitting the ground at a 30° angle spreads the same amount of light over twice as much area (if we imagine the sun shining from the south at noon, the north-south width doubles; the east-west width does not). Consequently, the amount of light falling on each square mile is only half as much.

This 'projection effect' is the main reason why the polar regions are much colder than equatorial regions on Earth. On an annual average the poles receive less insolation than does the equator, because at the poles the Earth's surface are angled away from the Sun.

Earth's Insolation

Direct insolation is the solar irradiance measured at a given location on Earth with a surface element perpendicular to the Sun's rays, excluding diffuse insolation (the solar radiation that is scattered or reflected by atmospheric components in the sky). Direct insolation is equal to the solar constant minus the atmospheric losses due to absorption and scattering. While the solar constant varies with the Earth-Sun distance and solar cycles, the losses depend on the time of day (length of light's path through the atmosphere depending on the Solar elevation angle), cloud cover, moisture content, and other impurities. Insolation is a fundamental abiotic factor affecting the metabolism of plants and the behaviour of animals.

Over the course of a year the average solar radiation arriving at the top of the Earth's atmosphere is roughly 1,366 watts per square metre. The radiant power is distributed across the entire electromagnetic spectrum, although most of the power is in the visible light portion of the spectrum. The Sun's rays are attenuated as they pass through the atmosphere, thus reducing the insolation at the Earth's surface to approximately 1,000 watts per square metre for a surface perpendicular to the Sun's rays at sea level on a clear day.

The actual figure varies with the Sun angle at different times of year, according to the distance the sunlight travels through the air, and depending on the extent of atmospheric haze and cloud cover. Ignoring clouds, the average insolation for the Earth is approximately 250 watts per square metre (6 (kW·h/m^2)/day), taking into account the lower radiation intensity in early morning and evening, and its near-absence at night.

The insolation of the sun can also be expressed in Suns, where one Sun equals 1,000 W/m^2 at the point of arrival, with kWh/(m^2·day) displayed as hours/day. When calculating the output of, for example, a photovoltaic panel, the angle of the sun relative to the panel needs to be taken into account as well as the insolation. (The insolation, taking into account the attenuation of the atmosphere, should be multiplied by the cosine of the angle between the normal to the panel and the direction of the sun from it). One Sun is a unit of power flux, not a standard value for actual insolation. Sometimes this unit is referred to as a Sol, not to be confused with a sol, meaning one solar day on, for example, a different planet, such as Mars.

Applications

In spacecraft design and planetology, it is the primary variable affecting equilibrium temperature.

In construction, insolation is an important consideration when designing a building for a particular climate. It is one of the most important climate variables for human comfort and building energy efficiency.

The projection effect can be used in architecture to design buildings that are cool in summer and warm in winter, by providing large vertical windows on the equator-facing side of the building (the south face in the northern hemisphere, or the north face in the southern hemisphere): this maximises insolation in the winter months when the Sun is low in the sky, and minimises it in the summer when the noonday Sun is high in the sky. (The Sun's north/south path through the sky spans 47 degrees through the year).

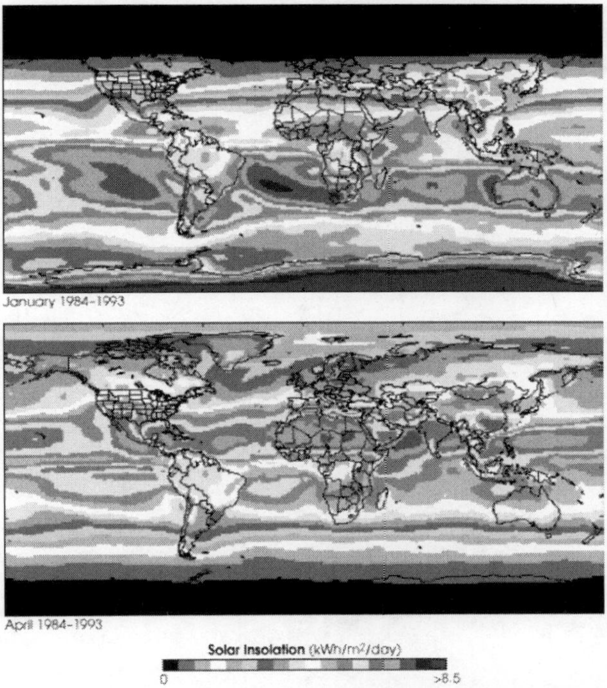

Figure: *Insolation variation by month; 1984-1993 averages for January (top) and April (bottom).*

Insolation figures are used as an input to worksheets to size solar power systems for the location where they will be installed. This can be misleading since insolation figures assume the panels are parallel with the ground, when in fact they are almost always mounted at an angle to face towards the sun. This gives inaccurately low estimates for winter. The figures can be obtained from an insolation map or by city or region from insolation tables that were generated with historical data over the last 30–50 years. Photovoltaic panels are rated under standard conditions to determine the Wp rating (watts peak), which can then be used with the insolation of a region to determine the expected output, along with other factors such as tilt, tracking and shading (which can be included to create the installed Wp rating). Insolation values range from 800 to 950 kWh/(kWp·y) in Norway to up to 2,900 in Australia.

In the fields of civil engineering and hydrology, numerical models of snowmelt runoff use observations of insolation. This permits estimation of the rate at which water is released from a melting snowpack. Field measurement is accomplished using a pyranometre.

Heat Budget of the Earth

The Earth can be considered as a physical system with an energy budget that includes all gains of incoming energy and all losses of outgoing energy.

The planet is approximately in equilibrium, so the sum of the gains is approximately equal to the sum of the losses. Note on accompanying images: These graphics depict only *net* energy transfer.

There is no attempt to depict the role of greenhouse gases and the exchange that occurs between the Earth's surface and the atmosphere or any other exchanges.

The Energy Budget

Incoming Energy

The total solar flux of energy entering the Earth's atmosphere is estimated at 174 petawatts. This flux consists of:

- solar radiation (99.97%, or nearly 173 petawatts; or about 340 W m^{-2})
  - This is equal to the product of the solar constant, about 1,366 watts per square metre, and the area of the Earth's disc as seen from the Sun, about 1.28×10^{14} square metres, averaged over the Earth's surface, which is four times larger. The solar flux averaged over just the sunlit half of the Earth's surface is about 680 W m^{-2}.
  - This is the incident energy. The energy actually absorbed by the earth is lower by a factor of the co-albedo; this is discussed in the next section.
  - Note that the solar constant varies (by approximately 0.1% over a solar cycle); and is not known absolutely to within better than about one watt per square metre.

Hence geothermal, tidal, and waste heat contributions are less uncertain than solar power.

- geothermal energy (0.025%; or about 44 to 47 terawatts; or about 0.08 W m^{-2})
  - — This is produced by stored heat and heat produced by radioactive decay leaking out of the Earth's interior.
- tidal energy (0.002%, or about 3 terawatts; or about 0.0059 W m^{-2})
  - — This is produced by the interaction of the Earth's mass with the gravitational fields of other bodies such as the Moon and Sun.
- waste heat from fossil fuel consumption (about 0.007%, or about 13 terawatts; or about 0.025 W m^{-2}) The total energy used by commercial energy sources from 1880 to 2000 (including fossil fuels and nuclear) is calculated to be 17.3 x 10^{21}Joules.

There are other minor sources of energy that are usually ignored in these calculations: accretion of interplanetary dust and solar wind, light from distant stars, the thermal radiation of space.

Although these are now known to be negligibly small, this was not always obvious: Joseph Fourier initially thought radiation from deep space was significant when he discussed the Earth's energy budget in a paper often cited as the first on the greenhouse effect.

Outgoing Energy

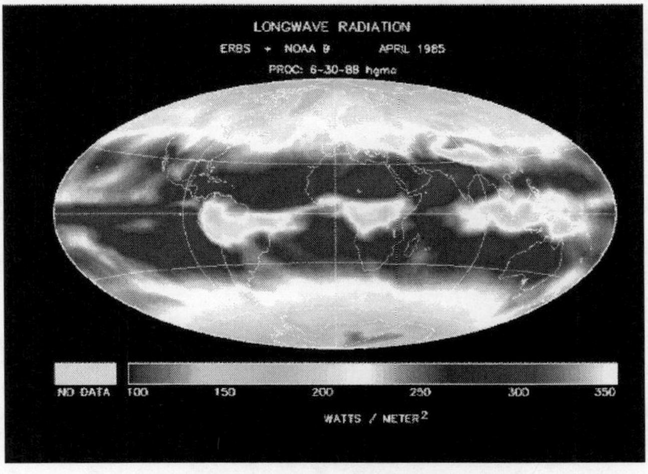

Figure: *Earth's longwave thermal radiation intensity, from clouds, atmosphere and ground.*

The average albedo (reflectivity) of the Earth is about 0.3, which means that 30% of the incident solar energy is reflected into space, while 70% is absorbed by the Earth and reradiated as infrared.

The planet's albedo varies from month to month and place to place, but 0.3 is the average figure. The contributions from geothermal and tidal power sources are so small that they are omitted from the following calculations.

30% of the incident energy is reflected, consisting of:

- 6% reflected from the atmosphere
- 20% reflected from clouds
- 4% reflected from the ground (including land, water and ice)

The remaining 70% of the incident energy is absorbed:

- 51% is absorbed by land and water, and then emerges in the following ways:
  - 23% is transferred back into the atmosphere as latent heat by the evaporation of water, called latent heat flux
  - 7% is transferred back into the atmosphere by heated rising air, called Sensible heat flux
  - 6% is radiated directly into space
  - 15% is transferred into the atmosphere by radiation, then reradiated into space
- 19% is absorbed by the atmosphere and clouds, including:
  - 16% reradiated into space
  - 3% transferred to clouds, from where it is radiated back into space

When the Earth is at thermal equilibrium, the same 70% that is absorbed is reradiated:

- 64% by the clouds and atmosphere
- 6% by the ground

Distribution of Temperature, Atmospheric Pressure and General Circulation of Winds

Local fire-weather elements-wind, temperature, moisture, and stability-respond continually to the varying patterns of pressure systems and to the changing properties of huge masses of air moving in generally predictable circulations over the earth's surface. These broad-scale circulations determine the regional patterns of rapidly changing fire weather-long term trends resulting in periods of wetness or drought and above or below-normal temperatures, and in seasonal changes in fire weather.

If we are to become acquainted with these variations in fire weather, we must understand how they are brought about, and the settings in which they take place.

The response to overall airflow applies also to local fuel conditions, so an understanding of general air circulation within the troposphere is essential to a usable knowledge of wild-land fire behaviour.

General Circulation

So far we have been concerned principally with the static properties of the atmosphere-its temperature, moisture, and pressure. In this chapter we will begin a more detailed consideration of the dynamics of the atmosphere-its motion-which was introduced.

We learned that the atmosphere is a gaseous mantle encasing the earth held there by gravity-and rotating with the earth. Within this huge envelope of air there are motions of a variable nature. If forces were not present to act on the atmosphere and upset its equilibrium, there would be no atmospheric motion-no circulation. The pressure exerted by the weight of the atmosphere would be the same everywhere at a given level. But disturbing forces are present. The earth is not heated uniformly, and the resultant unequal heating of the atmosphere causes compensating air motions, which tend to reduce the horizontal temperature differences.

The actual motions that are developed within the atmosphere are extremely complex and are not yet fully understood. Theories and models, which have been derived, are not wholly accepted because they do not completely account for all of the observed atmospheric motions. Most of the major features of the global circulations are rather well understood. Therefore, future modifications of present-day theories resulting from further research will not seriously affect out understanding of the general circulation as it relates to fire weather.

Primary Circulation

In equatorial regions the earth's surface receives more solar energy from the sun than it radiates back to space, and therefore acts as a heat source for the air in these regions. In polar regions the earth's surface radiates more energy into space than it receives from the sun. Since equatorial regions do not get hotter and hotter, and since polar regions do not become progressively colder, there must be some net transport of heat energy from equatorial to polar regions. Just how this is accomplished is one of the major features of the general circulation that is not completely understood. This transport could be accomplished by closed horizontal "cells" with north-south flow, by large-scale eddies, or, perhaps, by both methods. We will consider both methods.

Convective Circulation

Let us suppose that the earth's surface was uniform, that the earth did not rotate, and that it was uniformly heated around the entire Equator. Certainly this is a very hypothetical situation, but let us accept it for the sake of development of our discussion. We know that regions of warm surfaces, which warm the air overlying them, are characterised by rising air. The warming air expands and is forced aloft by the cooler, denser air flowing in from adjacent areas. In equatorial regions the warm air would rise to near the tropopause, reach a level of the same air density, then spread out and flow both north and south. As it moved toward the poles, it would cool by radiation and sink as its density increased. In the polar regions it would descend and begin to move toward the Equator.

In this hypothetical case the transport of heat could take take place by simple convective circulation. At the earth's surface there would be a permanent low-pressure belt around the earth at the Equator and a high-pressure area at each pole.

Since the earth does rotate, and since the sun is its single source of energy, this simple convective pattern cannot exist. The real circulation patterns are the result of the unequal heating mentioned above combined with the effect of the earth's rotation and the unequal partitioning of heat due to the uneven distribution of land and sea areas. Before we discuss the circulation on a rotating earth with a uniform surface, we will need to consider why and how the earth's rotation affects airflow.

How the Earth's Rotation Affects Airflow: Coriolis Force

If a maps of air, or any other body, moves in a straight line as viewed from a position in space, its path as viewed from a position on the earth is curved. The curvature indicates a deflection to the right in the Northern Hemisphere and a deflection to the left in the Southern Hemisphere. The reason for the deflection is that the earth, rotating toward the east on its axis, turns underneath the moving air or body. This deflective force is called the Coriolis force. It is an apparent rather than a real force, but since we are stationed on earth and view motions from the earth, the deflection is real from all earthbound positions.

To visualize the Coriolis force, let us consider a large disk or merry-go-round, rotating in a counterclockwise direction, as representing the Northern Hemisphere. A boy tossing a ball from the centre outward would find that the ball made a straight path in space, but traced a curved path on the disk below showing a deflection toward the right. Although more difficult to visualize, it is a fact that if the boy were stationed at any place on the rotating disk and tossed the ball in any horizontal direction, the ball would trace a curved path on the disk with a deflection to the right.

On the rotating earth, an air current in the Northern Hemisphere starting as a southerly wind, that is, moving toward the north, would be deflected to the right and become a southwest or west wind. Likewise, a north wind deflected to the right becomes a northeast or east wind. Since the northward airflow aloft just north of the equatorial region becomes nearly a true westerly flow, the northward movement is slowed and the air "piles up" at about latitude 30°N. The air also loses considerable heat by radiation.

Because of the piling up and the heat loss, some of the air descends, producing a surface high-pressure belt, while the rest continues in the westerly current aloft. Air that has descended flows both northward toward the pole and southward toward the Equator at the surface. Again, the effect of the earth's rotation comes into play. The northward-flowing current is turned to the right and becomes the prevailing westerlies of middle latitudes. The southward-flowing current, also deflected to the right, becomes the northeast trades of the low latitudes.

The air aloft that gradually moves northward continue to lose heat. In the polar regions it descends, gives up additional heat to the surface, and flows southward. This current is also turned to the right by the Coriolis force and becomes the polar easterlies of high latitudes. The cold air gradually pushes southward and finally meets the northward-flowing tropical air in what is referred to as the polar front zone. The polar and tropical air masses, which have different densities, tend to resist mixing. Instead, the lighter tropical air flows up and over the forward edge of the denser polar air.

This type of cellular circulation causes air to accumulate in the polar region. When the mass becomes great enough, the polar front zone is pushed southward, and the cold polar air penetrates to fairly low latitudes in a "polar outbreak". In this simplified circulation system, heat energy is carried northward by the airflow aloft, and cold air moves southward in cold outbreaks to maintain a balance of energy between equatorial and polar regions.

This primary circulation system results in the formation of several well-defined major regional circulation patterns or wind belts, some of which we have already mentioned.

These are known as: Doldrums, trade winds, horse latitudes, prevailing westerlies, polar front zone, and polar easterlies. The equatorial region of warm and moist rising air currents is referred to as the doldrums. It is a region of light surface winds, considerable cloudiness, and widespread shower activity. When the doldrum belt moves north from the Equator, as it does in the summer and early fail, it becomes the "breeding ground" for tropical storms and hurricanes.

Near latitude 30°N. is a region of descending air and high pressures known m the horse latitudes. As we will see later, the high atmospheric pressure in this region is usually best developed over the oceans. The high-pressure areas are characterised by light variable winds, little cloudiness, and little rainfall. Between the doldrums and the horse latitudes is the belt of trade winds-northeast trades in the Northern Hemisphere and southeast trades in the Southern Hemisphere.

The belt of westerlies extends from about 30°N. to about 55°N. North of here are the polar easterlies mentioned earlier. The polar front zone, between the prevailing westerlies and polar easterlies, is a zone of storminess, cloudiness, and precipitation. Its position around the hemisphere is extremely variable. Sometimes it plunges far southward into middle latitudes with cold air outbreaks; at other times it is carried far northward with intrusions of tropical air to high latitudes. We will see later that it is tied to the circulation aloft, particularly to the meandering stream of westerly winds in the upper troposphere.

Pressure Patterns

The simple primary circulation described above should result in a band of low pressure around the earth in the equatorial region, a band of high pressure about latitude 30°N., a band of low pressure in the polar front zone, and an area of high pressure in the polar region.

However, if we study the distribution of pressure over the Northern Hemisphere we do not find the bands to be entirely uniform. Instead we find pressure cells-areas with higher or lower pressure than the surrounding region.

Some of these are semi permanent cells, which remain relatively fixed; others are migratory. The weather is closely related to these pressure cells and other pressure patterns. If we are to understand and predict the weather, we need to determine the distribution of atmospheric pressure.

We learned that the atmosphere has mass and that atmospheric pressure is the result of the force of gravity acting on this mass.

Atmospheric pressure can be measured by balancing the weight of the atmosphere against that of a column of mercury. This is done with a mercurial barometre.

Another type of barometre, called an aneroid, has a partially vacuated metallic cell, so constructed that the sides tend to collapse under increasing atmospheric pressure and to expand with decreasing pressure. This movement is magnified by levers and is transmitted to a hand or pen, which indicates the pressure reading on a scale. We also learned in chapter I that atmospheric pressure at any location varies with time and decreases with altitude.

Constant-Level, Constant-Pressure Charts

To study the pressure distribution, we need, first of all, pressure measurements taken simultaneously at a number of stations. Meteorologists refer to these as synoptic observations.

Secondly, since stations are at different elevations and we wish to compare one pressure measurement with another, we need to correct the pressures to a common level, usually sea level. This is done by adding to the station pressure the weight of a hypothetical column of air extending from the level of the station down to sea level. Corrected readings are collected at a central point and plotted on a weather map. Such weather maps are called synoptic charts.

A graphical picture of the pressure distribution is obtained by drawing lines, called isobars, through points of equal pressure. Isobars are labelled in millibars (mb.) and are drawn usually for intervals of 4 mb., although the interval may vary with the map scale. Isobars may be thought of as contours of pressure, somewhat similar to contours of elevations on a topographic map. Pressure patterns aloft are also important in determining the structure of the atmosphere. They are portrayed in a slightly different way, however. Instead of determining pressure variations at a constant level, such as is done on the sea-level chart just described, the variations in the height of a constant-pressure surface are charted.

The pressure surfaces used in the 73 troposphere are 850 Tabs. (about 5,000 feet), 700 mb. (about 10,000 feet), 500 mb. (about 18,000 feet), and 300 mb. (about 30,000 feet). The heights above sea level, usually in tens of metres, of the pressure surface at a number of stations are plotted on a weather map. Contours of height for 60-metre intervals are drawn through points of equal height. These contours are strictly analogous to the contours on a topographic map. The only difference is that the constant pressure chart depicts the height of a pressure surface, while the topographic map depicts the height of the ground surface above sea level.

For our purpose, it makes little difference whether we think of pressure distribution in terms of a constant-level or constant-pressure chart. Areas of high pressure on a constant level chart would appear as areas of high heights on a constant-pressure chart, and low-pressure areas would show up as low heights.

Lows, Troughs

When a weather map is analyzed as described above, we find certain configurations, or patterns. On a sea-level chart we will find areas that have a lower pressure than the surrounding region. These are called low-pressure centres or areas, or simply Lows for short. They are also called cyclones because the air flows around them in a cyclonic direction (counterclockwise in the Northern Hemisphere). Lows are usually characterised by inward and rising air motion, which results in cooling and increased relative humidity. Sufficient lifting with adequate moisture will produce condensation of water vapor into clouds and may result in precipitation. Latent energy released by the condensation adds to the energy of the circulation system. A line of low pressure is referred to as a trough. The pressure along the line is lower than the pressure on either side. The isobars show a cyclonic curvature at the trough line but do not form a closed circulation. The characteristics of a trough are similar to those of a Low. Frequently a trough delineates the boundary between two different airflows and is a common place for the development of storm centres.

Highs, Ridges

High-pressure cells are another type of pressure pattern observed on analyzed weather maps. A high-pressure area is surrounded on all sides by lower pressure. We call it a High for short. It may also be referred to as an anticyclone because the windflaw around a High is anticyclone (clockwise in the Northern Hemisphere). The airflow in a High is generally outward and descending. For this reason, Highs are usually areas of minimum cloudiness and little or no precipitation. If the air descends from very high altitudes, it may be extremely dry. Ridges are lines of high pressure. The pressure is higher along the ridge than on either side. The curvature of isobars at a ridgeline is anticyclonic, but the isobars do not form a closed circulation. Ridges exhibit characteristics similar to Highs, with descending air and a minimum of cloudiness and precipitation.

Pressure and Wind Relations

Air always moves in response to pressure differences. As it moves, its speed and direction of motion are governed by a combination of forces. These include the pressure-gradient force, which causes air to move from high to low pressure; the Coriolis force, which causes a deflection to the right in the Northern Hemisphere; an outwardly directed centrifugal force if air is flowing in a curved path; and friction, which opposes all air movement near the surface of the earth.

Airflow can take place along a straight or curved path. Let us consider first the simpler case, that is, straight flow at a level high enough in the atmosphere so that friction with the earth's surface is negligible. For this case, only two of the forces mentioned above need be considered the pressure-gradient force and the Coriolis force.

Pressure gradient may be defined as the change of pressure per unit distance, for example, millibars per 100 miles. On the sea level map, as mentioned above, isobars are drawn for specific intervals of pressure. The closer the isobar spacing, the stronger the pressure gradient, and vice versa. The pressure gradient force tends to make air flow across the isobars from high to low pressure. But, as the air moves, it is deflected to the right by the Coriolis force. This force acts in a direction perpendicular to the airflow, and its magnitude depends upon both the speed of the airflow and upon the latitude. The reason for this is that the Coriolis force is caused by the rotation of the earth's surface beneath the airflow, and the rotation of the surface around a vertical axis depends upon the latitude. This rotation, and therefore the latitudinal effect of the Coriolis force, is greatest at the poles and decreases to zero at the Equator.

Geostrophic, Flow

Balance between the pressure-gradient force and the Coriolis force is achieved when these two forces oppose each other with equal magnitudes. The resulting flow is then parallel to the isobars, rather than across the isobars. If these forces are diagrammed, the pressure-gradient force is drawn at right angles to the isobars in the direction of low pressure. The Coriolis force is drawn at right angles to the line of motion and is directed toward the right in the Northern Hemisphere.

At a given latitude the speed of the airflow, increases with an increased pressure gradient - a decrease in the distance between isobars. With equal pressure gradients, a greater air speed will occur at lower latitudes than at higher latitudes because of the influence of latitude on the Coriolis force.

Straight flow, which occurs in regions of straight-line isobars, is termed geostrophic flow and implies balance between the pressure gradient and Coriolis forces.

Gradient Flow: Highs and Lows

In most areas on a weather map the isobars are curved rather than straight. The result is that as air moves, the direction of the pressure gradient force changes, and so does the airflow, to follow the curving isobars.

Here, an additional force must be considered-the outwardly directed centrifugal force. For steady motion, a balance must exist between the pressure-gradient force, the Coriolis force, and the centrifugal force. When these forces are in balance, the airflow is still parallel to the isobars, but it is known as gradient flow. As with geostrophic flow, high pressure is on the right in the Northern Hemisphere, as one looks downstream. Therefore, the direction of flow is always clockwise around a high-pressure centre and counterclockwise around a Low.

The balance of forces for gradient flow is more complicated than for geostrophic flow. If the forces around a Low were diagrammed, the pressure-gradient force is drawn at right angles to the isobars and directed inward. The Coriolis force is at right angles to the airflow and directed toward the right, which is outward, and the centrifugal force is at right angles to the isobars and directed outward. When the three forces are in balance; the pressure-gradient force balances the sum of the Coriolis and centrifugal forces. If the forces around a High are diagrammed, the centrifugal force is, of course, still directed outward. But now the pressure gradient force is directed outward, and the Coriolis force is directed inward. This means that the sum of the pressure-gradient and centrifugal forces balances the Coriolis force.

In both low- and high-pressure systems, the speed of the wind increases with increased pressure gradient; that is, with closer spacing of the isobars. Because the centrifugal force is added to the pressure-gradient force in a High, and subtracted from it in a Low, the wind speed in a High will be greater than in a Low with the same pressure gradient. In spite of this, we find higher wind speeds in Lows because the pressure gradients are usually much stronger.

One other characteristic difference also exists. In a low-pressure system, increased pressure gradients and increased air speeds may occur as the centre is approached. We can have, and of ten do observe, very strong wind speeds near the centre of Lows. In a High, however, because of the balance of forces there is a limiting value of wind speed that cannot be exceeded as the centre is approached. We find, therefore, that Highs have low wind speeds and weak pressure gradients near their centres.

Friction

So far we have considered straight flow and curved flow at levels high enough in the atmosphere so that the force of friction could be disregarded. But when we consider airflow near the ground, we must account for the friction force. The effect of friction on airflow is to retard the movement. Therefore, friction is a force acting in a direction opposite to the airflow. Since the Coriolis force varies with the wind speed, a reduction in the wind speed because of friction produces a reduction in the Coriolis force.

For steady motion there must be a balance among the pressure-gradient, centrifugal, Coriolis, and friction forces. The resulting balanced motion is a flow directed slightly across the isobars from high to low pressure. The amount of deviation depends upon the roughness of the terrain and will vary from 10 to 15 degrees over water to 25 to 45 degrees over land. The speed of the airflow is always lower with friction than without friction.

Friction assists in the transfer of air from high- to low-pressure areas. Because of friction, air flows spirally outward from a High and spirally inward around a Low near the surface. A person standing with his back to the wind has high pressure to his right and a little to the rear, and low-pressure on his left and a little forward. The effect of friction is, of course, greatest near the surface and decreases upward in the atmosphere. The depth of its influence varies directly with surface roughness and with atmospheric instability.

Generally, at altitudes higher than 2,000 feet above the surface, the effect of friction can be disregarded. Above this altitude, the airflow tends to be more nearly parallel to the isobars. The development of new pressure systems, and the intensification or decay of existing systems, as well as the migrations of these systems, cause many deviations in observed wind speeds and directions. Additional deviations develop because of local terrain. The combined effects of these influences can be seen by comparing the observed surface winds with the sea-level isobars on a surface weather map.

Circulation Patterns at Upper Levels

Our discussion of Highs, Lows, troughs, ridges, and the relationship between pressure and wind has been concerned primarily with the surface map, which is a constant-level map. The same terms and the same relationships apply to constant pressure charts used to portray the upper-air circulations described earlier. The balance of forces for airflow on a constant-pressure chart is similar to that on a constant-level chart. The only difference is that the pressure-gradient force is represented by the gradient of height of the constant-pressure surface. The friction force is disregarded on upper-air charts.

The circulation patterns in the middle and upper troposphere are quite different from those near the surface. They are less complicated because the effects of local heating of land and water, and of topography on air movements are greatly reduced. The major or large-scale hemispheric circulations are more in evidence. Troughs and ridges are common, but completely closed circulations-Highs and Lows-tend to decrease in frequency with altitude.

Circumpolar Westerlies

Except for a deep layer of easterly flow in equatorial regions, which reaches to the upper troposphere, the airflow aloft in the Northern Hemisphere consists of a broad belt of westerly winds extending from the subtropics to the polar regions. This belt of westerlies forms a large circumpolar vortex. An upper-air chart of the Northern Hemisphere will show that this is not a smooth circular vortex; instead, it is a meandering current forming waves of varying amplitude and wavelength. These horizontal waves appear as part of the pattern of an upper-air chart.

Meteorologists classify the waves into two categories: Long waves which usually number three to seven around the hemisphere, and short waves which are superimposed on the pattern of long waves. The long waves move slowly. They may drift eastward slowly; remain stationary for a number of days, and even retrograde on occasion. The westerly current in a long-wave ridge may go far to the north and allow tropical air to be carried to high latitudes. In a long-wave trough, the westerlies may go far to the south and allow cold polar air to reach low latitudes.

A persistent long-wave pattern plays, an important role in prolonged periods of abnormal weather. The region beneath a long-wave ridge is likely to experience clear, dry weather with above-normal temperatures.

The region beneath a long-wave trough is likely to have cloudy, wet weather with below-normal temperatures. Two types of long-wave patterns in the belt of westerlies are distinguished. One is a large-amplitude, short-wavelength pattern, called meridional.

It is effective in carrying tropical air to high latitudes and polar air to low latitudes. The other is a small-amplitude, long-wavelength pattern, called zonal, in which the principal movement of Highs and Lows in midlatitudes is west to east.

Short waves are smaller, rapidly moving oscillations, which proceed through the long wave pattern. They move northward around long-wave ridges and southward through long wave troughs. The speed of the short waves is usually slower than the wind speed aloft, indicating that the air moves through the waves. The short waves are associated with migratory Lows and Highs at the surface, and their movement is about the same speed as the surface systems.

Long waves cannot be shown by lines because the exact positions are usually obscured by short waves. Generally, what one sees are the short-wave troughs and ridges. The long-wave trough positions are usually identified by the place where short-wave troughs deepen. The same applies to long-wave ridge positions and short-wave ridges.

The migration of large-scale eddies - the Highs and Lows - is the second method of transporting excess heat away from lower latitudes (mentioned at the beginning of this chapter). The cyclonically rotating Lows in their travel from lower to higher latitudes on the east side of long-wave troughs are effective in pulling warm tropical air far north ahead of them and cold polar air far south behind them. The Lows eventually dissipate at high latitudes. Cold polar Highs moving south, usually on the west side of long-wave troughs, eventually merge with semi-permanent Highs in the horse latitudes. Thus, these large cyclonic and anticyclonic eddies are mechanisms by which warm air is transported northward and cold air is transported southward across the middle-latitude belt of westerlies. Closed circulations are sometimes found within the troughs and ridges aloft. Contours may indicate a closed High in a large-amplitude, long-wave ridge. Closed Lows may be found in long-wave troughs, and occasionally in short-wave troughs.

Jet Stream

Within the belt of westerlies there is often a core of very strong winds, called a jet stream. This fast-flowing river of air near the tropopause has wind speeds of 50 to 150 or 200 m.p.h. It is usually 100 to 400 miles wide and 3,000 to 7,000 feet deep.

When more than one jet stream occurs, the principal one is the polar-front jet stream associated with the surface polar front. It meanders in a wavelike pattern as part of the general westerly flow. Like the polar front, it is stronger in some places than others. It rarely encircles the entire hemisphere as a continuous river of air. More frequently, it is found in segments 1,000 to 3,000 miles long.

The north-south temperature gradient in the upper troposphere is concentrated in the jet-stream region. In fact, the jet stream is found only in those areas where a marked temperature gradient has developed. Below the jet, the region to the right is warm as one faces downstream, and the region to the left is cold. Above the jet stream, the warm and cold regions are reversed.

The mean position of the jet stream, and the belt of westerlies in which it is embedded, shifts south in the winter and north in the summer with the seasonal migration of the polar front. As it moves southward in the winter it also moves to higher altitudes and, on the average, its speed increases. The seasonal position of the jet stream is related to seasonal weather. During some summers its mean position may not be as far north as usual, and this position reflects summers that are cooler than normal. Similarly, during winters that are milder than normal, the jet stream does not move as far south as usual.

Although the polar jet stream is the primary one, other jet streams may exist high above surface fronts where the temperature contrast between air masses is sharp. A second jet stream south of the polar front jet is referred to as the subtropical jet.

Typical Circulation Patterns

The circulations that we observe are the combined results of the primary it secondary circulations, which, in turn, are produced by the uneven heating of the earth because of differences in latitude and in the distribution of land and water masses.

Semi-Permanent Centres

As mentioned earlier, the nonuniform character of the earth's surface results in cells of high pressure in the horse latitudes and cells of low pressure in the polar front zone, rather than continuous belts. Some high- and low-pressure systems appear so consistently in certain areas that they are considered semi-permanent and are given names.

Those of interest to us are the Pacific High in the Pacific, the Azores-Bermuda High in the Atlantic, the Aleutian Low in the Northern Pacific, and the Icelandic Low in the Northern Atlantic. These may be displaced from their normal positions occasionally, and at times portions; will break off and become migratory, especially the Lows. Usually though, these semi-permanent centres will remain stationary and quite strong for several days or weeks. The Highs tend to be more persistent than the Lows.

The strength of these cells varies with the season, and the development of other, less permanent cells is also a function of seasons.

In the summer, the oceans are colder than the land, and high-pressure centres are well developed over the oceans. Low pressure, due to stronger heating, is found over the continents. Over Southern Asia a semi-permanent Low develops in summer and a similar Low on a smaller scale is found in our Southwest. During the winter, the continents are colder than the oceans. A seasonal High develops in Siberia, and high pressure is common over North America. The semi-permanent Lows over the warmer oceans are well developed.

Let us consider the summer and winter patterns over North America and the adjacent means in more detail.

In the summer, because of the comparative warmth of the land, high pressure in the horse latitude belt is not frequently observed over the land. However, the Azores-Bermuda High often extends into the Southeastern States. The Pacific High and the Azores-Bermuda High are strong and rather far north as compared to their winter positions.

The Icelandic Low is weak. The Aleutian Low is not present in the Aleutian area, but low pressure is found over Northeastern Siberia.

The intense summer heat over the dry Southwest forms a low-pressure area known as the California Heat Low. Temperature contrasts between equatorial and polar regions are smaller in summer than in winter. Pressure gradients are weak, and the resulting air motion is slow compared to winter.

Aloft, the circumpolar vortex is small. This means that the belt of westerlies, the jet stream, and the polar front are far north. The westerlies an weak and confined to a relatively narrow band. The tracks of most surface Lows are also rather far north; these Lows usually travel eastward through Southern Canada or the Northern States.

A few travel northeastward through the Southern and Eastern States or along the Atlantic coast. The tracks of polar Highs me similarly far north.

The strong Azores-Bermuda High and Pacific High have a pronounced influence on summer weather in certain regions. The circulation around the western and of the Azores - Bermuda High (Bermuda High for short) brings warm, moist tropical air from the Atlantic and Gulf of Mexico into most of the Eastern and Central United States.

When this High extends far westward across the Gulf States, moisture from the Gulf is effectively cut off, and the East has hot, dry weather. Along the Pacific coast, the Pacific High blocks most Lows and forces them far to the north. The eastern end of the Pacific High is a region of subsiding air aloft. This subsiding air, which overlays a shallow layer of cool, moist air carried to the mast by northwest winds, produces a very stable condition and results in dry summer weather along the coast.

During winter, the Aleutian and Icelandic Lows are well developed. The Aleutian Low extends from the Aleutian Islands into the Gulf of Alaska, and much stormy weather and precipitation in the Western States are associated with the movement of this low-pressure system or segments of the main cell which break off and move south and southeast. The strong circulation around the Icelandic Low produces northerly winds and frigid weather in the eastern section of the continent. The Pacific and Bermuda Highs are weaker and displaced farther south in winter than in summer. Temperature contrasts between the tropics and polar regions are greater, and the wind circulations, both aloft and at the surface, are correspondingly stronger.

Aloft, the circumpolar vortex is large, extending to much lower latitudes. The belt of westerlies is broad. The mean position of the polar front is farther south than in the summer. The tracks of Highs and Lows vary considerably, but many take tracks that are much farther south in winter than in summer.

Due to the intense cooling of land areas, particularly at higher latitudes, many cold highpressure mass develop over the northern half of the continent. Periodically, these high-pres. Sure cells move southward, bringing polar or arctic air to the rest of the continent. Stormy weather is produced where these cold outbreaks meet warm, moist tropical air. The coldest Highs in North America come from the Hudson Bay region or Northwest Canada, while milder Highs move in from the Pacific as break off cells from the Pacific High. Another wintertime feature is the Great Basin High. Cool air masses from either Canada or the Northern Pacific move into the Great Basin and tend to stagnate in this intermountain area. Dry winds, warmed adiabatically as air flows from higher to lower elevations-including the cast winds of Washington and Oregon, north winds in northern California, and Santa Ana winds in southern California are associated with the track and positioning of the Great Basin High.

A migratory low-pressure cell, called the Colorado Low, often develops east of the central Rockies in winter. The circulation system of this Low usually intensifies as it moves to the northeast, reaching maximum development in the Great Lakes or St. Lawrence River area. This Low is usually accompanied by strong winds and rain or snow. The passage of the Low is followed by northerly winds and a cold high-pressure area from the north moving into the Great Plains or Great Lakes region. Lows that reach the west coast from the Pacific Ocean sometimes move intact over the mountains and continue in an easterly direction along a path curved toward the northeast. Frequently, however, the track of the Low is discontinuous. The Low fills on the west side of the mountains, then reforms on the east side and resumes its eastward movement.

Special Cyclonic Systems

Hurricanes, tornadoes, and waterspouts are special forms of low-pressure systems. Hurricanes cover a vast area and are quite deep. They originate over warm ocean water in the doldrums or in waves in the subtropical easterlies; and produce heavy precipitation and powerful winds. Large amounts of energy are released to feed these systems through the condensation of water vapor. A distinctive feature of these tropical storms is the virtually calm winds and comparatively clear skies at the centre, or eye, of the stem. Hurricanes first move toward the west in the easterly flow and later usually turn north and are caught up in the belt of westerlies. They then take on the characteristics of middle-latitude low-pressure systems. They lose intensity rapidly if they move over land because of the increased friction and the loss of the continuous supply of moisture.

Tornadoes and waterspots are small lowpressure cells in the form of intense spinning vertexes. When they occur, they are associated with severe thunderstorms. Winds near the centre of a tornado are commonly 100 to 200 m.p.h. and may exceed 400 m.p.h. The pressure near the centre is extremely low. The great destruction of these terrifying storms is due both to the high winds and the explosive effects of a sudden reduction in pressure as the tornado passes. Tornadoes range from 500 to 2,000 feet in diametre and travel over the ground with a speed of 20 to 40 m.p.h. Waterspouts that develop from the cloud downward are simply tornadoes occurring over the water. Waterspouts develop from the water upward, Usually they are not as intense as tornadoes are weak compared to tornadoes, and dissipate occurring over land. Other "fair weather" rapidly when they move inland.

Monsoons and Jet Streams

Monsoon is traditionally defined as a seasonal reversing wind accompanied by corresponding changes in precipitation, but is now used to describe seasonal changes in atmospheric circulation and precipitation associated with the asymmetric heating of land and sea. Usually, the term monsoon is used to refer to the rainy phase of a seasonally-changing pattern, although technically there is also a dry phase.

The major monsoon systems of the world consist of the West African and Asia-Australian monsoons. The inclusion of the North and South American monsoons with incomplete wind reversal has been debated. The term was first used in English in British India (now India, Bangladesh and Pakistan) and neighbouring countries to refer to the big seasonal winds blowing from the Bay of Bengal and Arabian Sea in the southwest bringing heavy rainfall to the area.

Etymology

The English *monsoon* came from Portuguese *monção*, ultimately from Arabic *mawsim* ("season"), "perhaps partly via early modern

Dutch *monsun*". The Arabic-origin word *mausam* is also the word for "weather" in Hindi, Urdu, and several other North Indian languages.

History

Strengthening of the Asian monsoon has been linked to the uplift of the Tibetan Plateau after the collision of the Indian sub-continent and Asia around 50 million years ago. Many geologists believe the monsoon first became strong around 8 million years ago based on records from the Arabian Sea and the record of wind-blown dust in the Loess Plateau of China.

More recently, plant fossils in China and new long-duration sediment records from the South China Sea led to a timing of the monsoon starting 15-20 million years ago and linked to early Tibetan uplift.

Testing of this hypothesis awaits deep ocean sampling by the Integrated Ocean Drilling Program. The monsoon has varied significantly in strength since this time, largely linked to global climate change, especially the cycle of the Pleistocene ice ages. Timing of the monsoon strengthening of the Indian Monsoon of around 5 million years ago was suggested due to an interval of closing of the Indonesian Seaway to cold thermocline waters passage from the Pacific to the Indian Ocean which is believed to have resulted in an increased sea surface temperature in the Indian Ocean, which increased gyral circulation and then caused an increased intensity of the monsoon.

Five episodes during the Quaternary at 2.22 Ma (PL-1), 1.83 Ma (PL-2), 0.68 Ma (PL-3), 0.45 Ma (PL-4) and 0.04 Ma (PL-5) were identified which showed a weakening of Leeuwin Current (LC). The weakening of the LC would have an effect on the sea surface temperature (SST) field in the Indian Ocean, as the Indonesian through flow generally warms the Indian Ocean. Thus these five intervals could probably be those of considerable lowering of SST in the Indian Ocean and would have influenced Indian monsoon intensity.

During the weak LC, there is the possibility of reduced intensity of the Indian winter monsoon and strong summer monsoon, because of change in the Indian Ocean dipole due to reduction in net heat input to the Indian Ocean through the Indonesian through flow.

Thus a better understanding of the possible links between El Niño, Western Pacific Warm Pool, Indonesian Throughflow, wind pattern off western Australia, and ice volume expansion and contraction can be obtained by studying the behaviour of the LC during Quaternary at close stratigraphic intervals.

Process

Monsoons may be considered as large-scale sea breezes, due to seasonal heating and the resulting development of a thermal low over a continental landmass. They are caused by the larger amplitude of the seasonal cycle of land temperature compared to that of nearby oceans.

This differential warming happens because heat in the ocean is mixed vertically through a "mixed layer" that may be fifty metres deep, through the action of wind and buoyancy-generated turbulence, whereas the land surface conducts heat slowly, with the seasonal signal penetrating perhaps a metre or so. Additionally, the specific heat capacity of liquid water is significantly higher than that of most materials that make up land.

Together, these factors mean that the heat capacity of the layer participating in the seasonal cycle is much larger over the oceans than over land, with the consequence that the air over the land warms faster and reaches a higher temperature than the air over the ocean.

The hot air over the land tends to rise, creating an area of low pressure. This creates a steady wind blowing toward the land, bringing the moist near-surface air over the oceans with it. Similar rainfall is caused by the moist ocean air being lifted upwards by mountains, surface heating, convergence at the surface, divergence aloft, or from storm-produced outflows at the surface. However the lifting occurs, the air cools due to expansion in lower pressure, which in turn produces condensation.

In winter, the land cools off quickly, but the ocean retains heat longer. The cold air over the land creates a high pressure area which produces a breeze from land to ocean. Monsoons are similar to sea and land breezes, a term usually referring to the localized, diurnal (daily) cycle of circulation near coastlines, but they are much larger in scale, stronger and seasonal.

Most summer monsoons have a dominant westerly component and a strong tendency to ascend and produce copious amounts of rain (because of the condensation of water vapor in the rising air). The intensity and duration, however, are not uniform from year to year. Winter monsoons, by contrast, have a dominant easterly component and a strong tendency to diverge, subside and cause drought.

Even more broadly, it is now understood that in the geological past, monsoon systems likely accompanied the formation of supercontinents such as Pangaea, with their extreme continental climates.

5

Nature of Cartography

Cartography is the study and practice of making maps (also can be called mapping). Combining science, aesthetics, and technique, cartography builds on the premise that reality can be modeled in ways that communicate spatial information effectively.

The fundamental problems of traditional cartography are to:

- Set the map's agenda and select traits of the object to be mapped. This is the concern of map editing. Traits may be physical, such as roads or land masses, or may be abstract, such as toponyms or political boundaries.
- Represent the terrain of the mapped object on flat media. This is the concern of map projections.
- Eliminate characteristics of the mapped object that are not relevant to the map's purpose. This is the concern of generalization.
- Reduce the complexity of the characteristics that will be mapped. This is also the concern of generalization.
- Orchestrate the elements of the map to best convey its message to its audience. This is the concern of map design.

Modern cartography is closely integrated with geographic information science (GIScience) and constitutes many theoretical and practical foundations of geographic information systems.

History

The earliest known map is a matter of some debate, both because the definition of "map" is not sharp and because some artifacts speculated to be maps might actually be something else. A wall painting, which may depict the ancient Anatolian city of Catalhoyuk (previously known as Catal Huyuk or Catal Huyuk), has been dated to the late 7th millennium BCE. Other known maps of the ancient world include the Minoan "House of the Admiral" wall painting from c. 1600 BCE, showing a seaside community in an oblique perspective and an engraved map of the holy Babylonian city of Nippur, from the Kassite period (14th – 12th centuries BCE). The oldest surviving world maps are the Babylonian world maps from the 9th century BCE. One shows Babylon on the Euphrates, surrounded by a circular landmass showing Assyria, Urartu and several cities, in turn surrounded by a "bitter river" (Oceanus), with seven islands arranged around it. Another depicts Babylon as being further north from the centre of the world.

The ancient Greeks and Romans created maps, beginning at latest with Anaximander in the 6th century BC. In the 2nd century AD, Ptolemy produced his treatise on cartography, Geographia. This contained Ptolemy's world map - the world then known to Western society *(Ecumene)*. As early as the 700s, Arab scholars were translating the works of the Greek geographers into Arabic.

In ancient China, geographical literature spans back to the 5th century BC. The oldest extant Chinese maps come from the State of Qin, dated back to the 4th century BC, during the Warring States Period. In the book of the *Xin Yi Xiang Fa Yao*, published in 1092 by the Chinese scientist Su Song, a star map on the equidistant cylindrical projection. Although this method of charting seems to have existed in China even prior to this publication and scientist, the greatest significance of the star maps by Su Song is that they represent the oldest existent star maps in printed form.

Early forms of cartography of India included legendary paintings; maps of locations described in Indian epic poetry, for example, the *Ramayana*. Indian cartographic traditions also covered the locations of the Pole star, and other constellations of use. These charts may have been in use by the beginning of the Common Era for purposes of navigation.

Mappa mundi is the general term used to describe Medieval European maps of the world. Approximately 1,100 mappae mundi are known to have survived from the Middle Ages. Of these, some 900 are found illustrating manuscripts and the remainder exist as stand-alone documents (Woodward).

The Arab geographer, Muhammad al-Idrisi, produced his medieval atlas *Tabula Rogeriana* in 1154. He incorporated the knowledge of

Africa, the Indian Ocean and the Far East, gathered by Arab merchants and explorers with the information inherited from the classical geographers to create the most accurate map of the world up until his time. It remained the most accurate world map for the next three centuries.

In the Age of Exploration, from the 15th century to the 17th century, European cartographers both copied earlier maps (some of which had been passed down for centuries) and drew their own based on explorers' observations and new surveying techniques. The invention of the magnetic compass, telescope and sextant enabled increasing accuracy. In 1492, Martin Behaim, a German cartographer, made the oldest extant globe of the Earth.

Johannes Werner refined and promoted the *Werner map projection*. In 1507, Martin Waldseemüller produced a globular world map and a large 12-panel world wall map (*Universalis Cosmographia*) bearing the first use of the name "America". Portuguese cartographer, Diego Ribero, was author of the first known planisphere with a graduated Equator (1527). Italian cartogapher Battista Agnese produced at least 71 manuscript atlases of sea charts.

Due to the sheer physical difficulties inherent in cartography, map-makers frequently lifted material from earlier works without giving credit to the original cartographer. For example, one of the most famous early maps of North America is unofficially known as the "Beaver Map", published in 1715 by Herman Moll. This map is an exact reproduction of a 1698 work by Nicolas de Fer. De Fer in turn had copied images that were first printed in books by Louis Hennepin, published in 1697, and Francois Du Creux, in 1664. By the 1700s, map-makers started to give credit to the original engraver by printing the phrase "After [the original cartographer]" on the work.

Technological Changes

In cartography, technology has continually changed in order to meet the demands of new generations of mapmakers and map users. The first maps were manually constructed with brushes and parchment; therefore, varied in quality and were limited in distribution. The advent of magnetic devices, such as the compass and much later, magnetic storage devices, allowed for the creation of far more accurate maps and the ability to store and manipulate them digitally.

Advances in mechanical devices such as the printing press, quadrant and vernier, allowed for the mass production of maps and the ability to make accurate reproductions from more accurate data. Optical technology, such as the telescope, sextant and other devices that use telescopes, allowed for accurate surveying of land and the ability of mapmakers and navigators to find their latitude by measuring angles to the North Star at night or the sun at noon.

Advances in photochemical technology, such as the lithographic and photochemical processes, have allowed for the creation of maps that have fine details, do not distort in shape and resist moisture and wear. This also eliminated the need for engraving, which further shortened the time it takes to make and reproduce maps.

Advances in electronic technology in the 20th century ushered in another revolution in cartography. Ready availability of computers and peripherals such as monitors, plotters, printers, scanners (remote and document) and analytic stereo plotters, along with computer programs for visualization, image processing, spatial analysis, and database management, have democratized and greatly expanded the making of maps. The ability to superimpose spatially located variables onto existing maps created new uses for maps and new industries to explore and exploit these potentials.

These days most commercial-quality maps are made using software that falls into one of three main types: CAD, GIS and specialized illustration software. Spatial information can be stored in a database, from which it can be extracted on demand. These tools lead to increasingly dynamic, interactive maps that can be manipulated digitally.

Map Types

General vs Thematic Cartography

In understanding basic maps, the field of cartography can be divided into two general categories: general cartography and thematic cartography. General cartography involves those maps that are constructed for a general audience and thus contain a variety of features. General maps exhibit many reference and location systems and often are produced in a series. For example, the 1:24,000 scale topographic maps of the United States Geological Survey (USGS) are a standard as compared to the 1:50,000 scale Canadian maps. The government of the UK produces the classic 1:50,000 (replacing the older 1 inch to 1 mile) "Ordnance Survey" maps of the entire UK and with a range of correlated larger- and smaller-scale maps of great detail.

Thematic cartography involves maps of specific geographic themes, oriented toward specific audiences. A couple of examples might be a dot map showing corn production in Indiana or a shaded area map of Ohio counties, divided into numerical choropleth classes. As the volume of geographic data has exploded over the last century, thematic cartography has become increasingly useful and necessary to interpret spatial, cultural and social data.

An orienteering map combines both general and thematic cartography, designed for a very specific user community. The most prominent thematic element is shading, that indicates degrees of difficulty of travel due to vegetation. The vegetation itself is not identified, merely classified by the difficulty ("fight") that it presents.

Topographic vs Topological

A topographic map is primarily concerned with the topographic description of a place, including (especially in the 20th century) the use of contour lines showing elevation. Terrain or relief can be shown in a variety of ways.

A topological map is a very general type of map, the kind you might sketch on a napkin. It often disregards scale and detail in the interest of clarity of communicating specific route or relational information. Beck's London Underground map is an iconic example. Though the most widely used map of "The Tube," it preserves little of reality: It varies scale constantly and abruptly, it straightens curved tracks, and it contorts directions haphazardly. The only topography on it is the River Thames, letting the reader know whether a station is north or south of the river. That and the topology of station order and interchanges between train lines are all that is left of the geographic space. Yet those are all a typical passenger wishes to know, so the map fulfils its purpose.

Map Design

Map Purpose and Informations' Selection

Arthur H. Robinson, an American cartographer influential in thematic cartography, stated that a map not properly designed "will be a cartographic failure." He also claimed, when considering all aspects of cartography, that "map design is perhaps the most complex." Robinson codified the mapmaker's understanding that a map must be designed foremost with consideration to the audience and its needs.

From the very beginning of mapmaking, maps "have been made for some particular purpose or set of purposes". The intent of the map should be illustrated in a manner in which the percipient acknowledges its purpose in a timely fashion. The term *percipient* refers to the person receiving information and was coined by Robinson. The principle of figure-ground refers to this notion of engaging the user by presenting a clear presentation, leaving no confusion concerning the purpose of the map. This will enhance the user's experience and keep his attention. If the user is unable to identify what is being demonstrated in a reasonable fashion, the map may be regarded as useless.

Making a meaningful map is the ultimate goal. Alan MacEachren explains that a well designed map "is convincing because it implies authenticity" (1994). An interesting map will no doubt engage a reader. Information richness or a map that is multivariate shows relationships within the map. Showing several variables allows comparison, which adds to the meaningfulness of the map. This also generates hypothesis and stimulates ideas and perhaps further research. In order to convey the message of the map, the creator must design it in a manner which will aid the reader in the overall understanding of its purpose. The title of a map may provide the "needed link" necessary for communicating that message, but the overall design of the map fosters the manner in which the reader interprets it (Monmonier, 1993).

In the 21st century it is possible to find a map of virtually anything from the inner workings of the human body to the virtual worlds of cyberspace. Therefore there are now a huge variety of different styles and types of map - for example, one area which has evolved a specific and recognisable variation are those used by public transport organisations to guide passengers, namely urban rail and metro maps, many of which are loosely based on 45 degree angles as originally perfected by Harry Beck and George Dow.

Naming Conventions

Most maps use text to label places and for such things as a map title, legend and other information. Maps are often made in specific languages, though names of places often differ between languages. So a map made in English may use the name *Germany* for that country, while a German map would use *Deutschland* and a French map *Allemagne*. A word that describes a place, using a non-native terminology or language is referred to as an exonym.

In some cases the proper name is not clear. For example, the nation of Burma officially changed its name to Myanmar, but many nations do not recognize the ruling junta and continue to use *Burma*. Sometimes an official name change is resisted in other languages and the older name may remain in common use. Examples include the use of *Saigon* for Ho Chi Minh City, *Bangkok* for Krung Thep and *Ivory Coast* for Cote d'Ivoire.

Difficulties arise, when transliteration or transcription between writing systems is required. National names tend to have well established names in other languages and writing systems, such as *Russia*, but for many placenames a system of transliteration or transcription is required. In transliteration, the symbols of one language are represented by symbols in another. For example, the Cyrillic letter *P* is traditionally written as *R* in the Latin alphabet. Systems exist for transliteration of Arabic, but the results may vary. For example, the Yemeni city of Mocha is written variously in English as Mocha, Al Mukha, al-Mukha, Mocca and Moka. Transliteration systems are based on relating written symbols to one another, while transcription is the attempt to spell in one language the phonetic sounds of another. Chinese writing is transformed into the Latin alphabet through the Pinyin phonetic transcription systems. Other systems were used in the past, such as Wade-Giles, resulting in the city being spelled *Beijing* on newer English maps and *Peking* on older ones.

Further difficulties arise when countries, especially former colonies, do not have a strong national geographic naming standard. In such cases, cartographers may have to choose between various phonetic spellings of local names versus older imposed, sometimes resented, colonial names. Some countries have multiple official languages, resulting in multiple official placenames. For example, the capital of Belgium is both *Brussels* and *Bruxelles*. In Canada, English and French are official languages and places have names in both languages. British Columbia is also officially named *la Colombie-Britannique*. English maps rarely show the French names outside of Quebec, which itself is spelled *Quebec* in French. The study of placenames is called toponymy, while that of the origin and historical usage of placenames as words is etymology.

In order to improve legibility or to aid the illiterate, some maps have been produced using pictograms to represent places. The iconic example of this practice is Lance Wyman's early plans for the Mexico City Metro, on which stations were shown simply as stylized logos. Wyman also prototyped such a map for the Washington Metro, though ultimately the idea was rejected. Other cities experimenting with such maps are Fukuoka, Guadalajara and Monterrey.

Map Symbology

The quality of a map's design affects its reader's ability to extract information and to learn from the map. Cartographic symbology has been developed in an effort to portray the world accurately and effectively convey information to the map reader. A legend explains the pictorial language of the map, known as its symbology.

The title indicates the region the map portrays; the map image portrays the region and so on. Although every map element serves some purpose, convention only dictates inclusion of some elements, while others are considered optional. A menu of map elements includes the neatline (border), compass rose or north arrow, overview map, bar scale, projection and information about the map sources, accuracy and publication.

When examining a landscape, scale can be intuited from trees, houses and cars. Not so with a map. Even such a simple thing as a north arrow is crucial. It may seem obvious that the top of a map should point north, but this might not be the case.

Colour, likewise, is equally important. How the cartographer displays the data in different hues can greatly affect the understanding or feel of the map. Different intensities of hue portray different objectives the cartographer is attempting to get across to the audience. Today, personal computers can display up to 16 million distinct colours at a time, even though the human eye can distinguish only a minimum number of these (Jeer, 1997).

This fact allows for a multitude of colour options for even for the most demanding maps. Moreover, computers can easily hatch patterns in colours to give even more options. This is very beneficial, when symbolizing data in categories such as quintile and equal interval classifications.

Quantitative symbols give a visual measure of the relative size/importance/number that a symbol represents and to symbolize this data on a map, there are two major classes of symbols used for portraying quantitative properties. Proportional symbols change their visual weight according to a quantitative property. These are appropriate for extensive statistics. Choropleth maps portray data collection areas, such as counties or census tracts, with colour. Using colour this way, the darkness and intensity (or value) of the colour is evaluated by the eye as a measure of intensity or concentration (Harvard Graduate School of Design, 2005).

Map Generalization

A good map has to provide a compromise between portraying the items of interest (or themes) in the *right place* for the map scale used, against the need to annotate that item with text or a symbol, which takes up space on the map medium and very likely will cause some other item of interest to be displaced. The cartographer is thus constantly making judgements about what to include, what to leave out and what to show in a *slightly* incorrect place - because of the demands of the annotation. This issue assumes more importance as the scale of the map gets smaller (i.e. the map shows a larger area), because relatively, the annotation on the map takes up more space *on the ground*. A good example from the late 1980s was the Ordnance Survey's first digital maps, where the *absolute* positions of major roads shown at scales of 1:1250 and 1:2500 were sometimes a scale distance of hundreds of metres away from ground truth, when shown on digital maps at scales of 1:250000 and 1:625000, because of the overriding need to annotate the features.

Cartographic Errors

Some maps contain deliberate errors or distortions, either as propaganda or as a "watermark" helping the copyright owner identify infringement if the error appears in competitors' maps. The latter often come in the form of nonexistent, misnamed, or misspelled "trap streets". Other names and forms for this are paper townsites, fictitious entries, and copyright easter eggs.

Another motive for deliberate errors is simply cartographic graffiti or prank: a mapmaker wishing to leave his or her mark on the work. Mount Richard, for example, was a fictitious peak on the Rocky Mountains' continental divide that appeared on a Boulder County, Colorado map in the early 1970s. It is believed to be the work of drafts man Richard Ciacci. The fiction was not discovered until two years later.

Geography and Regionalism

Region is most commonly a geographical term that is used in various ways among the different branches of geography. In general, a region may be seen as a collection of smaller units (as in "the New England states") or as one part of a larger whole (as in "the New England region of the United States"). Regions can be defined by physical characteristics, human characteristics, and functional characteristics. As a way of describing spatial areas, the concept of regions is important and widely used among the many branches of geography, each of which can describe areas in regional terms. For example, ecoregion is a term used in environmental geography, cultural region in cultural geography, bioregion in biogeography, and so on. The field of geography that studies regions themselves is called regional geography.

In the fields of physical geography, ecology, biogeography, zoogeography, and environmental geography, regions tend to be based on natural features such as ecosystems or biotopes, biomes, drainage basins, mountain ranges, soil types.

Physiographic Regions

Regions defined on the basis of landform characteristics are called "physiographic" or "geomorphic" regions. Physiography involves the delineation and description of regions from the viewpoint of geomorphology. Geologist Nevin Fenneman defined a classic three-level hierarchical system of physiographic regions for the United States in 1946. The regions are called divisions, provinces, and sections. For example, there are 8 large physiographic divisions, such as the Canadian Shield and the Interior Plains. These are subdivioned into provinces and sections. The Appalachian Highlands division, for example, contains the Valley and Ridge province, which in turn contains three sections, the Tennessee section, the Middle section, and the Hudson section. The Valley and Ridge province approximately corresponds to the more general region known as the Ridge-and-valley Appalachians.

Palaeogeographic Regions

Palaeogeography is the study of ancient geologic environments. Since the physical structures of the Earth's surface have changed over geologic time, palaeogeographers have coined various names for ancient regions that no longer exist, from very large regions such as the supercontinents Rodinia, Pangaea, and Pannotia, to relatively small regions like Beringia. Other examples include the Tethys Ocean and Ancylus Lake. Palaeogeographic continental regions that include Laurentia, Proto-Laurasia, Laurasia, Euramerica (the "Old Red Continent"), and Gondwana. The Paleogeographic region is also where paleontologist find answers in history.

Historical Regions

The field of historical geography involves the study of human history as it relates to places and regions, or, inversely, the study of how places and regions have changed over time. D. W. Meinig, a historical geographer of America, describes many historical regions in his book *The Shaping of America: A Geographical Perspective on 500 Years of History*. For example, in identifying European "source regions" in early American colonization efforts, he defines and describes the "Northwest European Atlantic Protestant Region", which includes sub-regions such as the "Western Channel Community", which itself is made of sub-regions such as the "English West Country" of Cornwall, Devon, Somerset and Dorset.

In describing historic regions of America, Meinig writes of "The Great Fishery" off the coast of Newfoundland and New England, an oceanic region that includes the Grand Banks. He rejects regions traditionally used in describing American history, like New France, "West Indies", the Middle Colonies, and the individual colonies themselves (Province of Maryland, for example). Instead he writes of "discrete colonization areas", which may be named after colonies, but rarely adhere strictly to political boundaries. Historic regions of this type Meinig writes about include "Greater New England" and its major sub-regions of "Plymouth", "New Haven shores" (including parts of Long Island), "Rhode Island" (or "Narragansett Bay"), "the Piscataqua", "Massachusetts Bay", "Connecticut Valley", and to a lesser degree, regions in the sphere of influence of Greater New England, "Acadia" (Nova Scotia), "Newfoundland and The Fishery/ The Banks".

Other examples of historical regions include Iroquoia, Ohio Country, Illinois Country, and Rupert's Land.

Tourism Region

A tourism region is a geographical region that has been designated by a governmental organization or tourism bureau as having common cultural or environmental characteristics. These regions are often named after a geographical, former, or current administrative region or may have a name created for tourism purposes. The names often evoke certain positive qualities of the area and suggest a coherent tourism experience to visitors. Countries, states, provinces, and other administrative regions are often carved up into tourism regions which, in addition to drawing the attention of potential tourists, often provide tourists who are otherwise unfamiliar with an area with a manageable number of attractive options.

Some of the more famous tourism regions based on historical or current administrative regions include Tuscany in Italy and Yucatan in Mexico. Famous examples of regions created by a government or tourism bureau include the United Kingdom's Lake District and California's Wine Country great plains region.

Natural Resource Regions

Natural resources often occur in distinct regions. Natural resource regions can be a topic of physical geography or environmental geography, but also have a strong element of human geography and economic geography. A coal region, for example, is a physical or geomorphological region, but its development and exploitation can make it into an economic and a cultural region. Some examples of natural resource regions include the Rumaila Field, the oil field that lies along the border or Iraq and Kuwait and played a role in the Gulf War; the Coal Region of Pennsylvania, which is a historical region as well as a cultural, physical, and natural resource region; the South Wales Coalfield, which like Pennsylvania's coal region is a historical, cultural, and natural region; the Kuznetsk Basin, a similarly important coal mining region in Russia; Kryvbas, the economic and iron ore mining region of Ukraine; and the James Bay Project, a large region of Quebec where one of the largest hydroelectric systems in the world has been developed.

Religious Regions

Sometimes a region associated with a religion is given a name, like Christendom, a term with medieval and renaissance connotations of Christianity as a sort of social and political polity. The term Muslim world is sometimes used to refer to the region of the world where Islam is dominant. These broad terms are very vague when used to describe regions.

Within some religions there are clearly defined regions. The Roman Catholic Church, the Church of England, the Eastern Orthodox Church, and others, define ecclesiastical regions with names such as diocese, eparchy, ecclesiastical provinces, and parish.

For example, the United States is divided into 32 Roman Catholic ecclesiastical provinces. The Lutheran Church - Missouri Synod is organized into 33 geographic "districts", which are subdivided into "circuits" (the Atlantic District (LCMS), for example). The Church of Jesus Christ of Latter-day Saints uses regions similar to dioceses and parishes, but uses terms like ward and stake.

Political Regions

In the field of political geography regions tend to be based on political units such as sovereign states; subnational units such as provinces, counties, townships, territories, etc; and multinational groupings, including formally defined units such as the European Union, the Association of Southeast Asian Nations, and NATO, as well as informally defined regions such as the Third World, Western Europe, and the Middle East.

Local Administrative Regions

There are many relatively small regions based on local government agencies such as districts, agencies, or regions. In general, they are all regions in the general sense of being bounded spatial units. Examples include electoral districts such as Washington's 6th congressional district and Tennessee's 1st congressional district; school districts such as Granite School District and Los Angeles Unified School District; economic districts such as the Reedy Creek Improvement District; metropolitan areas such as the Seattle metropolitan area, and metropolitan districts such as the Metropolitan Water Reclamation District of Greater Chicago, the Las Vegas-Clark County Library District, the Metropolitan Police Service of Greater London, as well as other local districts like the York Rural Sanitary District, the Delaware River Port Authority, the Nassau County Soil and Water Conservation District, and C-TRAN.

Regional Government in Connecticut

In the U.S. State of Connecticut the roles of county governments are now performed by regional governments not abiding to the present county borders. Ever since the dissolution of county government in Connecticut in 1960, the roles of regional services once provided by the county are now provided by regional agencies of towns. Counties still are used in Connecticut as geographical entities and in some counties they are still used to organize judicial districts, also counties are still used to organize the state marshal system in Connecticut. Counties were also used to organize the sheriff's department of each Connecticut county until 2000, when county sheriff's were eliminated due to mismanagement as was the reason for abolishing the county governments. An example of one former county sheriff's department is the Fairfield County Sheriff's Department which served Fairfield County in Connecticut. All sheriff's departments in Connecticut were not eliminated, only at the county level. Several towns and cities in Connecticut still maintain a sheriff's department such as Shelton with the Shelton Sheriff's Department.

Administrative Regions

The word "region" is taken from the Latin *regio*, and a number of countries have borrowed the term as the formal name for a type of subnational entity (e.g., the *region*, used in Chile). In English, the word is also used as the conventional translation for equivalent terms in other languages.

In Spain the official name of the autonomous community of Murcia is *Region de Murcia*. Also, some single-province autonomous communities such as Madrid use the term *region* interchangeably with *comunidad autonoma*.

Two lan (counties) in Sweden are officially called 'regions': Skane and Vastra Gotaland, and there is currently a controversial proposal to divide the rest of Sweden into large regions, replacing the current counties.

The government of the Philippines uses the term "region" (in Filipino, *rehiyon*) when it's necessary to group provinces, the primary administrative subdivision of the country. This is also the case in Brazil which groups its primary administrative divisions (*estados*; "states") into *grandes regioes* (greater regions) for statistical purposes, while Russia uses (economic regions) in a similar way, as does Romania and Venezuela.

Traditional or Informal Regions

The traditional territorial divisions of some countries are also commonly rendered in English as "regions". These informal divisions do not form the basis of the modern administrative divisions of these countries, but still define and delimit local regional identity and sense of belonging. Examples include:

- Finland
- Japan
- Korea
- Norway (*landsdeler*)
- Romania
- Slovakia.

Geographical Regions

A region can also be used for a geographical area; with this usage, there is an implied distinctiveness about the area that defines it. Such a distinction is often made on the basis of a historical, political, or cultural cohesiveness that separates the region from its neighbours.

Geographical regions can be found within a country (e.g., the Midlands, in England), or transnationally (e.g., the Middle East).

Similarly, the United Nations Statistics Division has devised a scheme for classifying macrogeographic regions (continents), continental subregions, and selected socioeconomic groupings.

Functional Region

A functional region or Nodal region, is a region that has a defined core that retains a specific characteristic that diminishes outwards. To be considered a Functional region, at least one form of spatial interaction must occur between the centre and all other parts of the region. A functional region is organized around a node or focal point with the surrounding areas linked to that node by transportation systems, communication systems, or other economic association involving such activities as manufacturing and retail trading. A typical functional region is a metropolitan area (MA) as defined by the Bureau of Census. For example, the New York MA is a functional region that covers parts of several states. It is linked by commuting patterns, trade flows, television and radio broadcasts, newspapers, travel for recreation and entertainment. Other functional regions include shopping regions centred on malls or supermarkets, area served by branch banks, and ports and their hinterlands.

Military Regions

In military usage a region is shorthand for the name of a military formation larger than an Army Group and smaller than an Army Theater or simply Theater. The full name of the military formation is Army Region. An Army Region usually consists of between two and five Army Groups. The size of an Army Region can vary widely but is generally somewhere between about 1 million and 3 million soldiers. Two or more Army Regions could make up an Army Theater. An Army Region would typically be commanded by a full General (US four stars), a Field Marshal or General of the Army (US five stars), or Generalissimo (Soviet Union). Due to the large size of this formation, its use is rarely employed. Some of the very few examples of an Army Region would be each of the Eastern, Western, and southern (mostly in Italy) fronts in Europe during World War II. The military map unit symbol for this echelon of formation consists of six Xs.

6

The Geographical Distribution of Man

Human Geography

Human geography is one of the two major subfields of geography. Human geography is the study of human use and understanding of the Earth and the process which have affected this. It is linked to both social science and the humanities.

Human geography broadly differs from physical geography in that it has a greater focus on studying intangible or abstract patterns surrounding human activity and is more receptive to qualitative research methodologies. It encompasses human, political, cultural, social and economic aspects of the social sciences. While the major focus of human geography is not the physical landscape of the Earth, it is not possible to discuss human geography without going into the physical landscape with which human activities are being played out and environmental geography which is an important link between the two. Human geography is both methodologically and theoretically diverse, including feminist, marxist, post-structural approaches, among others, and using both qualitative methods (such as ethnographies and interviews) and quantitative methods (such as survey research, statistical analysis and model building).

Fields of Human Geography

The main fields of study in human geography focus around the core fields of:

Culture: Cultural geography is the study of cultural products and norms and their variation across and relations to spaces and places. It focuses on describing and analyzing the ways language, religion, economy, government, and other cultural phenomena vary or remain constant from one place to another and on explaining how humans function spatially.

Development: Development Geography is the study of the Earth's geography with reference to the Standard of living and the Quality of life of its human inhabitants, study of the location, distribution and spatial organization of economic activities across the Earth. The subject matter investigated is strongly influenced by the researcher's methodological approach.

Economic: Economic geography examines relationships between human economic systems, states, and other actors, and the biophysical environment.

Health: Health geography is the application of geographical information, perspectives, and methods to the study of health, disease, and health care.

Historical: Historical Geography is the study of the human, physical, fictional, theoretical, and "real" geographies of the past. Historical geography studies a wide variety of issues and topics. A common theme is the study of the geographies of the past and how a place or region changes through time. Many historical geographers study geographical patterns through time, including how people have interacted with their environment, and created the cultural landscape.

Political: Political geography is concerned with the study of both the spatially uneven outcomes of political processes and the ways in which political processes are themselves affected by spatial structures.

Population: Population geography is the study of the ways in which spatial variations in the distribution, composition, migration, and growth of populations are related to the nature of places.

Tourism: Tourism geography is the study of travel and tourism as an industry, as a human activity, and especially as a place-based experience.

Urban: Urban geography is the study of urban areas with specific regards to spatial and relational aspects and theories. That is the study of areas which have a high concentration of buildings and infrastructure. These are areas where the majority of economic activities are in the secondary sector and tertiary sectors. They probably have a high population density.

Human Migration

Human migration is physical movement by humans from one area to another, sometimes over long distances or in large groups. The movement of populations in modern times has continued under the form of both voluntary migration within one's region, country, or beyond, and involuntary migration (which includes the slave trade, Human traffic in human beings and ethnic cleansing). People who migrate are called migrants or more specifically, emigrants, immigrants, or settlers, depending on historical setting, circumstances and perspective.

The pressures of human migrations, whether as outright conquest or by slow cultural infiltration and resettlement, have affected the grand epochs in history and in land (for example, the Decline of the Roman Empire); under the form of colonization, migration has transformed the world (such as the prehistoric and historic settlements of Australia and the Americas). Population genetics studied in traditionally settled modern populations have opened a window into the historical patterns of migrations, a technique pioneered by Luigi Luca Cavalli-Sforza.

Forced migration has been a means of social control under authoritarian regimes yet free initiative migration is a powerful factor in social adjustment and the growth of urban populations.

In December 2003, The Global Commission on International Migration (GCIM) was launched with the support of Secretary-General of the United Nations Kofi Annan and several countries, with an independent 19-member Commission, a threefold mandate and a finite life-span ending December 2005. Its report, based on regional consultation meetings with stakeholders and scientific reports from leading international migration experts, was published and presented to Kofi Annan on 5 October 2005.

Different types of migration include:

- Seasonal human migration mainly related to agriculture.
- Rural to Urban, more common in developing countries as industrialization takes effect (urbanization)
- Urban to Rural, more common in developed countries due to a higher cost of urban living (suburbanization)
- International migration.

Pre-modern Migrations

Historical migration of human populations begins with the movement of *Homo erectus* out of Africa across Eurasia about a million years ago. *Homo sapiens* appear to have occupied all of Africa about 150,000 years ago, moved out of Africa 70,000 years ago, and had spread across Australia, Asia and Europe by 40,000 years BCE. Migration to the Americas took place 20,000 to 15,000 years ago, and by 2,000 years ago, most of the Pacific Islands were colonized. Later population movements notably include the Neolithic Revolution, Indo-European expansion, and the Early Medieval Great Migrations including Turkic expansion.

Early humans migrated due to many factors such as changing climate and landscape and inadequate food supply. The evidence indicates that the ancestors of the Austronesian peoples spread from the South Chinese mainland to Taiwan at some time around 8,000 years ago. Evidence from historical linguistics suggests that it is from this island that seafaring peoples migrated, perhaps in distinct waves separated by millennia, to the entire region encompassed by the Austronesian languages. It is believed that this migration began around 6,000 years ago. Indo-Aryan migration to and within Northern India is presumed to have taken place in the Middle to Late Bronze Age, contemporary to the Late Harappan phase in India (ca. 1700 to 1300 BC). From 180 BC, a series of invasions from Central Asia followed, including those led by the Indo-Greeks, Indo-Scythians, Indo-Parthians and Kushans in the north-western Indian subcontinent.

From about 750 BC, the Greeks began 250 years of expansion, settling colonies in all directions. In Europe two waves of migrations dominate demographic distributions, that of the Celtic people, and the later Migration Period from the east. Other examples are small movements like ancient Scots moving from Hibernia to Caledonia and Magyars into Pannonia (modern-day Hungary). Turkic peoples spread across most of Central Asia into Europe and the Middle East between the 6th and 11th centuries. Recent research suggests that Madagascar was uninhabited until Austronesian seafarers from Indonesia arrived during the 5th and 6th centuries AD. Subsequent migrations from both the Pacific and Africa further consolidated this original mixture, and Malagasy people emerged.

Before the expansion of the Bantu languages and their speakers, the southern half of Africa is believed to have been populated by Pygmies and Khoisan speaking people, today occupying the arid regions around the Kalahari Desert and the forest of Central Africa. By about 1000 AD Bantu migration had reached modern day Zimbabwe and South Africa. The Banu Hilal and Banu Maqil were a collection of Arab Bedouin tribes from the Arabian Peninsula who migrated westwards via Egypt between the 11th and 13th centuries. Their migration strongly contributed to the arabization and islamization of the western Maghreb, which was until then dominated by Berber tribes. Ostsiedlung was the medieval eastward migration and settlement of Germans. The 13th century was the time of the great Mongol and Turkic migrations across Eurasia.

Between the 11th and 18th centuries, the Vietnamese expanded southward in a process known as nam tien (southward expansion). Manchuria was separated from China proper by the Inner Willow Palisade, which restricted the movement of the Han Chinese into Manchuria during the Qing Dynasty, as the area was off-limits to the Han until the Qing started colonizing the area with them later on in the dynasty's rule.

The Age of Exploration and European Colonialism led to an accelerated pace of migration since Early Modern times. In the 16th century perhaps 240,000 Europeans entered American ports. In the 19th century over 50 million people left Europe for the Americas. The local populations or tribes, such as the Aboriginal people in Canada, Brazil, Argentina, Australia, Japan and the United States, were usually far overwhelmed numerically by the settlers. More recent examples are the movement of ethnic Chinese into Tibet and Xinjiang, ethnic Javanese into Western New Guinea and Kalimantan, Brazilians into Amazonia, Israelis into the West Bank and Gaza, ethnic Arabs into Iraqi Kurdistan, and ethnic Russians into Siberia and Central Asia.

Modern Migrations

Industrialization

While the pace of migration had accelerated since the 18th century already (including the involuntary slave trade), it would increase further in the 19th century. Manning distinguishes three major types of migration: labour migration, refugee migrations, and urbanization. Millions of agricultural workers left the countryside and moved to the cities causing unprecedented levels of urbanization. This phenomenon began in Britain in the late 18th century and spread around the world and continues to this day in many areas.

Industrialization encouraged migration wherever it appeared. The increasingly global economy globalized the labour market. The Atlantic slave trade diminished sharply after 1820, which gave rise to self-bound contract labour migration from Europe and Asia to plantations. Overpopulation, open agricultural frontiers, and rising industrial centres attracted voluntary migrants. Moreover, migration was significantly made easier by improved transportation techniques.

Transnational labour migration reached a peak of three million migrants per year in the early twentieth century. Italy, Norway, Ireland and the Quongdong region of China were regions with especially high emigration rates during these years. These large migration flows influenced the process of nation state formation in many ways. Immigration restrictions have been developed, as well as diaspora cultures and myths that reflect the importance of migration to the foundation of certain nations, like the American melting pot. The transnational labour migration fell to a lower level from 1930s to the 1960s and then rebounded.

The United States experienced considerable internal migration related to industrialization, including its African American population. From 1910–1970, approximately 7 million African Americans migrated from the rural Southern United States, where blacks faced both poor economic opportunities and considerable political and social prejudice, to the industrial cities of the Northeast, Midwest and West where relatively well paid jobs were available. This phenomenon came to be known in the United States as its own Great Migration. With the demise of legalized segregation in the 1960s and greatly improved economic opportunities in the South in the subsequent decades, millions of blacks have returned to the South from other parts of the country since 1980 in what has been called the New Great Migration.

The twentieth century experienced also an increase in migratory flows caused by war and politics. Muslims moved from the Balkan to Turkey, while Christians moved the other way, during the collapse of the Ottoman Empire. 400,000 Jews moved to Palestine in the early twentieth century. The Russian Civil War caused some 3 million Russians, Poles and Germans to migrate out of the Soviet Union. World War II and decolonization also caused migrations.

World War II

The Jewish communities across Europe, the Mediterranean and the Middle East were formed from voluntary and involuntary migrants. After the Holocaust (1938 to 1945), there was increased migration to the British Mandate of Palestine, which became the modern state of Israel as a result of the United Nations Partition Plan for Palestine.

Provisions of the Potsdam Agreement from 1945 signed by victorious Western Allies and the Soviet Union led to one of the largest European migrations, and the largest in the 20th century. It involved the migration and resettlement of close to or over 20 million people.

The largest affected group were 16.5 million Germans expelled from Eastern Europe westwards. The second largest group were Poles, millions of whom were expelled westwards from eastern Kresy region and resettled in the so-called Recovered Territories. Hundreds of thousands of Poles, Ukrainians (Operation Vistula), Lithuanians, Latvians, Estonians and some Belarusians, were expelled eastwards from Europe to the Soviet Union. Finally, many of the several hundred thousand Jews remaining in Eastern Europe after the Holocaust migrated outside Europe to Israel and the United States.

India

In 1947, upon the Partition of India, large populations moved from India to Pakistan and vice versa, depending on their religious beliefs. The partition was promulgated in the Indian Independence Act 1947 as a result of the dissolution of the British Indian Empire. The partition displaced up to 12.5 million people in the former British Indian Empire, with estimates of loss of life varying from several hundred thousand to a million.

Muslim residents of the former British India migrated to Pakistan (including East Pakistan which is now Bangladesh), whilst Hindu and Sikh residents of Pakistan and Hindu residents of East Pakistan (now Bangladesh) moved in the opposite direction.

In modern India, estimates based on industry sectors mainly employing migrants suggest that there are around 100 million circular migrants in India. Caste, social networks and historical precedents play a powerful role in shaping patterns of migration. Migration for the poor is mainly circular, as despite moving temporarily to urban areas, they lack the social security which might keep them there more permanently. They are also keen to maintain a foothold in home areas during the agricultural season.

Research by the Overseas Development Institute identifies a rapid movement of labour from slower to faster growing parts of the economy. Migrants can often find themselves excluded by urban housing policies and migrant support initiatives are needed to give workers improved access to market information, certification of identity, housing and education.

Theories

Ravenstein

Certain laws of social science have been proposed to describe human migration. The following was a standard list after Ravenstein's proposals during the time frame of 1834 to 1913. The laws are as follows:

1. every migration flow generates a return or countermigration.
2. the majority of migrants move a short distance.
3. migrants who move longer distances tend to choose big-city destinations
4. urban residents are often less migratory than inhabitants of rural areas.
5. families are less likely to make international moves than young adults;
   - Migration stage by stage
   - Urban Rural difference
   - Migration and Technology
   - Economic condition.

Lee

Lee's Push-pull theory divides factors causing migrations into two groups of factors: Push and pull factors. Push factors are things that are bad about the country that one lives in and pull factors are things that attract one to another area.

Push Factors:

- Not enough jobs
- Few opportunities
- Primitive conditions
- Desertification
- Famine or drought
- Political fear or persecution
- Poor medical care
- Loss of wealth
- Natural disasters
- Death threats
- Lack of political or religious freedom

- Pollution
- Poor housing
- Landlord/tenant issues
- Bullying
- Discrimination
- Poor chances of marrying.

Pull Factors:

- Job opportunities
- Better living conditions
- Political and/or religious freedom
- Enjoyment
- Education
- Better medical care
- Attractive climates
- Security
- Family links
- Industry
- Better chances of marrying.

Climate Cycles

The modern field of climate history suggests that the successive waves of Eurasian nomadic movement throughout history have had their origins in climatic cycles, which have expanded or contracted pastureland in Central Asia, especially Mongolia and the Altai. People were displaced from their home ground by other tribes trying to find land that could be grazed by essential flocks, each group pushing the next further to the south and west, into the highlands of Anatolia, the plains of Hungary, into Mesopotamia or southwards, into the rich pastures of China.

Other Models

Migration occurs because individuals search for food, sex and security outside their usual habitation. (Idyorough, 2008) Idyorough is of the view that towns and cities are a creation of the human struggle to obtain food, sex and security. Migration occurs because individuals search for food, sex and security outside their usual habitation. To produce food, security and human reproduction of its species, human beings must, out of necessity, move out of their usual habitation and enter into indispensable social relationships that are cooperative or antagonistic. Human beings also develop tools/equipment to enable them interact with nature to produce the desired food and security.

The improved relationship (cooperative relationships) among human beings and improved technology further conditioned by the push and pull factors all interact together to cause or bring about migration and higher concentration of individuals into towns and cities. The higher the technology of production of food and security and the higher the cooperative relationship among human beings in the production of food and security and in the reproduction of the human species the higher would be the push and pull factors in the migration and concentration of human beings in towns and cities. Countryside, towns and cities do not just exist but they do so to meet the human basic needs of food, security and the reproduction of the human species. Therefore, migration occurs because individuals search for food, sex and security outside their usual habitation. Social services in the towns and cities are provided to meet these basic needs for human survival and pleasure.

Nature

Nature, in the broadest sense, is equivalent to the natural world, physical world, or material world. "Nature" refers to the phenomena of the physical world, and also to life in general. It ranges in scale from the subatomic to the cosmic.

The word *nature* is derived from the Latin word *natura,* or "essential qualities, innate disposition", and in ancient times, literally meant "birth". *Natura* was a Latin translation of the Greek word *physis,* which originally related to the intrinsic characteristics that plants, animals, and other features of the world develop of their own accord. The concept of nature as a whole, the physical universe, is one of several expansions of the original notion; it began with certain core applications of the word by pre-Socratic philosophers, and has steadily gained currency ever since. This usage was confirmed during the advent of modern scientific method in the last several centuries.

Within the various uses of the word today, "nature" may refer to the general realm of various types of living plants and animals, and in some cases to the processes associated with inanimate objects– the way that particular types of things exist and change of their own accord, such as the weather and geology of the Earth, and the matter and energy of which all these things are composed. It is often taken to mean the "natural environment" or wilderness–wild animals, rocks, forest, beaches, and in general those things that have not been substantially altered by human intervention, or which persist despite human intervention. For, example, manufactured objects and human interaction generally are not considered part of nature, unless qualified as, for example, "human nature" or "the whole of nature". This more traditional concept of natural things which can still be found today implies a distinction between the natural and the artificial, with the artificial being understood as that which has been brought into being by a human consciousness or a human mind. Depending on the particular context, the term "natural" might also be distinguished from the unnatural, the supernatural, and the artifactual.

Earth

Earth (or, "the earth") is the only planet presently known to support life, and its natural features are the subject of many fields of scientific research. Within the solar system, it is third nearest to the sun; it is the largest terrestrial planet and the fifth largest overall. Its most prominent climatic features are its two large polar regions, two relatively narrow temperate zones, and a wide equatorial tropical to subtropical region. Precipitation varies widely with location, from several metres of water per year to less than a millimetre. 71 percent of the Earth's surface is covered by salt-water oceans. The remainder consists of continents and islands, with most of the inhabited land in the Northern Hemisphere.

Earth has evolved through geological and biological processes that have left traces of the original conditions. The outer surface is divided into several gradually migrating tectonic plates, which have changed relatively quickly several times. The interior remains active, with a thick layer of molten mantle and an iron-filled core that generates a magnetic field.

The atmospheric conditions have been significantly altered from the original conditions by the presence of life-forms, which create an ecological balance that stabilizes the surface conditions. Despite the wide regional variations in climate by latitude and other geographic factors, the long-term average global climate is quite stable during interglacial periods, and variations of a degree or two of average global temperature have historically had major effects on the ecological balance, and on the actual geography of the Earth.

Geology

Geology is the science and study of the solid and liquid matter that constitutes the Earth. The field of geology encompasses the study of the composition, structure, physical properties, dynamics, and history of Earth materials, and the processes by which they are formed, moved, and changed. The field is a major academic discipline, and is also important for mineral and hydrocarbon extraction, knowledge about and mitigation of natural hazards, some engineering fields, and understanding past climates and environments.

Geological Evolution

The geology of an area evolves through time as rock units are deposited and inserted and deformational processes change their shapes and locations.

Rock units are first emplaced either by deposition onto the surface or intrude into the overlying rock. Deposition can occur when sediments settle onto the surface of the Earth and later lithify into sedimentary rock, or when as volcanic material such as volcanic ash or lava flows, blanket the surface. Igneous intrusions such as batholiths, laccoliths, dikes, and sills, push upwards into the overlying rock, and crystallize as they intrude.

After the initial sequence of rocks has been deposited, the rock units can be deformed and/or metamorphosed. Deformation typically occurs as a result of horizontal shortening, horizontal extension, or side-to-side (strike-slip) motion. These structural regimes broadly relate to convergent boundaries, divergent boundaries, and transform boundaries, respectively, between tectonic plates.

Historical Perspective

Earth is estimated to have formed 4.54 billion years ago from the solar nebula, along with the Sun and other planets. The moon formed roughly 20 million years later. Initially molten, the outer layer of the planet cooled, resulting in the solid crust. Outgassing and volcanic activity produced the primordial atmosphere. Condensing water vapour, most or all of which came from ice delivered by comets, produced the oceans and other water sources. The highly energetic chemistry is believed to have produced a self-replicating molecule around 4 billion years ago.

Continents formed, then broke up and reformed as the surface of Earth reshaped over hundreds of millions of years, occasionally combining to make a supercontinent. Roughly 750 million years ago, the earliest known supercontinent Rodinia, began to break apart. The continents later recombined to form Pannotia which broke apart about 540 million years ago, then finally Pangaea, which broke apart about 180 million years ago.

There is significant evidence that a severe glacial action during the Neoproterozoic era covered much of the planet in a sheet of ice. This hypothesis has been termed the "Snowball Earth", and it is of particular interest as it precedes the Cambrian explosion in which multicellular life forms began to proliferate about 530–540 million years ago.

Since the Cambrian explosion there have been five distinctly identifiable mass extinctions. The last mass extinction occurred some 65 million years ago, when a meteorite collision probably triggered the extinction of the non-avian dinosaurs and other large reptiles, but spared small animals such as mammals, which then resembled shrews. Over the past 65 million years, mammalian life diversified.

Several million years ago, a species of small African ape gained the ability to stand upright. The subsequent advent of human life, and the development of agriculture and further civilization allowed humans to affect the Earth more rapidly than any previous life form, affecting both the nature and quantity of other organisms as well as global climate. By comparison, the oxygen catastrophe, produced by the proliferation of algae during the Siderian period, required about 300 million years to culminate.

The present era is classified as part of a mass extinction event, the Holocene extinction event, the fastest ever to have occurred. Some, such as E. O. Wilson of Harvard University, predict that human destruction of the biosphere could cause the extinction of one-half of all species in the next 100 years. The extent of the current extinction event is still being researched, debated and calculated by biologists.

Atmosphere, Climate, and Weather

The atmosphere of the Earth serves as a key factor in sustaining the planetary ecosystem. The thin layer of gases that envelops the Earth is held in place by the planet's gravity. Dry air consists of 78% nitrogen, 21% oxygen, 1% argon and other inert gases, carbon dioxide, etc.; but air also contains a variable amount of water vapour. The atmospheric pressure declines steadily with altitude, and has a scale height of about 8 kilometres at the Earth's surface: the height at which the atmospheric pressure has declined by a factor of *e* (a mathematical constant equal to 2.71...). The ozone layer of the Earth's atmosphere plays an important role in depleting the amount of ultraviolet (UV) radiation that reaches the surface. As DNA is readily damaged by UV light, this serves to protect life at the surface. The atmosphere also retains heat during the night, thereby reducing the daily temperature extremes.

Terrestrial weather occurs almost exclusively in the lower part of the atmosphere, and serves as a convective system for redistributing heat. Ocean currents are another important factor in determining climate, particularly the major underwater thermohaline circulation which distributes heat energy from the equatorial oceans to the polar regions. These currents help to moderate the differences in temperature between winter and summer in the temperate zones. Also, without the redistributions of heat energy by the ocean currents and atmosphere, the tropics would be much hotter, and the polar regions much colder.

Weather can have both beneficial and harmful effects. Extremes in weather, such as tornadoes or hurricanes and cyclones, can expend large amounts of energy along their paths, and produce devastation. Surface vegetation has evolved a dependence on the seasonal variation of the weather, and sudden changes lasting only a few years can have a dramatic effect, both on the vegetation and on the animals which depend on its growth for their food.

The planetary climate is a measure of the long-term trends in the weather. Various factors are known to influence the climate, including ocean currents, surface albedo, greenhouse gases, variations in the solar luminosity, and changes to the planet's orbit. Based on historical records, the Earth is known to have undergone drastic climate changes in the past, including ice ages.

The climate of a region depends on a number of factors, especially latitude. A latitudinal band of the surface with similar climatic attributes forms a climate region. There are a number of such regions, ranging from the tropical climate at the equator to the polar climate in the northern and southern extremes. Weather is also influenced by the seasons, which result from the Earth's axis being tilted relative to its orbital plane. Thus, at any given time during the summer or winter, one part of the planet is more directly exposed to the rays of the sun. This exposure alternates as the Earth revolves in its orbit. At any given time, regardless of season, the northern and southern hemispheres experience opposite seasons. Weather is a chaotic system that is readily modified by small changes to the environment, so accurate weather forecasting is currently limited to only a few days. Overall, two things are currently happening worldwide: (1) temperature is increasing on the average; and (2) regional climates have been undergoing noticeable changes.

Water on Earth

Water is a chemical substance that is composed of hydrogen and oxygen and is vital for all known forms of life. In typical usage, *water* refers only to its liquid form or state, but the substance also has a solid state, ice, and a gaseous state, water vapour or steam.

Water covers 71% of the Earth's surface. On Earth, it is found mostly in oceans and other large water bodies, with 1.6% of water below ground in aquifers and 0.001% in the air as vapour, clouds (formed of solid and liquid water particles suspended in air), and precipitation. Oceans hold 97% of surface water, glaciers and polar ice caps 2.4%, and other land surface water such as rivers, lakes and ponds 0.6%. Additionally, a minute amount of the Earth's water is contained within biological bodies and manufactured products.

Oceans

An ocean is a major body of saline water, and a principal component of the hydrosphere. Approximately 71% of the Earth's surface (an area of some 361 million square kilometres) is covered by ocean, a continuous body of water that is customarily divided into several principal oceans and smaller seas. More than half of this area is over 3,000 meters (9,800 ft) deep. Average oceanic salinity is around 35 parts per thousand (ppt) (3.5%), and nearly all seawater has a salinity in the range of 30 to 38 ppt. Though generally recognized as several 'separate' oceans, these waters comprise one global, interconnected body of salt water often referred to as the World Ocean or global ocean. This concept of a global ocean as a continuous body of water with relatively free interchange among its parts is of fundamental importance to oceanography.

The major oceanic divisions are defined in part by the continents, various archipelagos, and other criteria: these divisions are (in descending order of size) the Pacific Ocean, the Atlantic Ocean, the Indian Ocean, the Southern Ocean and the Arctic Ocean. Smaller regions of the oceans are called seas, gulfs, bays and other names. There are also salt lakes, which are smaller bodies of landlocked saltwater that are not interconnected with the World Ocean. Two notable examples of salt lakes are the Aral Sea and the Great Salt Lake.

Lakes

A lake (from Latin *lacus*) is a terrain feature (or physical feature), a body of liquid on the surface of a world that is localized to the bottom of basin (another type of landform or terrain feature; that is, it is not global) and moves slowly if it moves at all. On Earth, a body of water is considered a lake when it is inland, not part of the ocean, is larger and deeper than a pond, and is fed by a river. The only world other than Earth known to harbour lakes is Titan, Saturn's largest moon, which has lakes of ethane, most likely mixed with methane. It is not known if Titan's lakes are fed by rivers, though Titan's surface is carved by numerous river beds. Natural lakes on Earth are generally found in mountainous areas, rift zones, and areas with ongoing or recent glaciation. Other lakes are found in endorheic basins or along the courses of mature rivers. In some parts of the world, there are many lakes because of chaotic drainage patterns left over from the last Ice Age. All lakes are temporary over geologic time scales, as they will slowly fill in with sediments or spill out of the basin containing them.

Ponds

A pond is a body of standing water, either natural or man-made, that is usually smaller than a lake. A wide variety of man-made bodies of water are classified as ponds, including water gardens designed for aesthetic ornamentation, fish ponds designed for commercial fish breeding, and solar ponds designed to store thermal energy. Ponds and lakes are distinguished from streams via current speed. While currents in streams are easily observed, ponds and lakes possess thermally driven microcurrents and moderate wind driven currents. These features distinguish a pond from many other aquatic terrain features, such as stream pools and tide pools.

Rivers

A river is a natural watercourse, usually freshwater, flowing toward an ocean, a lake, a sea or another river. In a few cases, a river simply flows into the ground or dries up completely before reaching another body of water. Small rivers may also be called by several other names, including stream, creek, brook, rivulet, and rill; there is no general rule that defines what can be called a river. Many names for small rivers are specific to geographic location; one example is *Burn* in Scotland and North-east England. Sometimes a river is said to be larger than a creek, but this is not always the case, due to vagueness in the language. A river is part of the hydrological cycle. Water within a river is generally collected from precipitation through surface runoff, groundwater recharge, springs, and the release of stored water in natural ice and snowpacks (i.e., from glaciers).

Streams

A stream is a flowing body of water with a current, confined within a bed and stream banks. In the United States a stream is classified as a watercourse less than 60 feet (18 metres) wide. Streams are important as conduits in the water cycle, instruments in groundwater recharge, and they serve as corridors for fish and wildlife migration. The biological habitat in the immediate vicinity of a stream is called a riparian zone. Given the status of the ongoing Holocene extinction, streams play an important corridor role in connecting fragmented habitats and thus in conserving biodiversity. The study of streams and waterways in general is known as *surface hydrology* and is a core element of environmental geography.

Ecosystems

Ecosystems are composed of a variety of abiotic and biotic components that function in an interrelated way. The structure and composition is determined by various environmental factors that are interrelated. Variations of these factors will initiate dynamic modifications to the ecosystem. Some of the more important components are: soil, atmosphere, radiation from the sun, water, and living organisms.

Central to the ecosystem concept is the idea that living organisms interact with every other element in their local environment. Eugene Odum, a founder of ecology, stated: “Any unit that includes all of the organisms (i.e.: the “community”) in a given area interacting with the physical environment so that a flow of energy leads to clearly defined trophic structure, biotic diversity, and material cycles (i.e.: exchange of materials between living and nonliving parts) within the system is an ecosystem.” Within the ecosystem, species are connected and dependent upon one another in the food chain, and exchange energy and matter between themselves as well as with their environment. The human ecosystem concept is grounded in the deconstruction of the human/nature dichotomy and the premise that all species are ecologically integrated with each other, as well as with the abiotic constituents of their biotope.

A smaller unit of size is called a microecosystem. For example, a microsystem can be a stone and all the life under it. A *macroecosystem* might involve a whole ecoregion, with its drainage basin.

Wilderness

Wilderness is generally defined as areas that have not been significantly modified by human activity. The WILD Foundation goes into more detail, defining wilderness as: "The most intact, undisturbed wild natural areas left on our planet - those last truly wild places that humans do not control and have not developed with roads, pipelines or other industrial infrastructure." Wilderness areas can be found in preserves, estates, farms, conservation preserves, ranches, National Forests, National Parks and even in urban areas along rivers, gulches or otherwise undeveloped areas. Wilderness areas and protected parks are considered important for the survival of certain species, ecological studies, conservation, solitude, and recreation. Some nature writers believe wilderness areas are vital for the human spirit and creativity, and some Ecologists consider wilderness areas to be an integral part of the planet's self-sustaining natural ecosystem (the biosphere). They may also preserve historic genetic traits and that they provide habitat for wild flora and fauna that may be difficult to recreate in zoos, arboretums or laboratories.

Life

Although there is no universal agreement on the definition of life, scientists generally accept that the biological manifestation of life is characterized by organization, metabolism, growth, adaptation, response to stimuli and reproduction. Life may also be said to be simply the characteristic state of organisms.

Properties common to terrestrial organisms (plants, animals, fungi, protists, archaea and bacteria) are that they are cellular, carbon-and-water-based with complex organization, having a metabolism, a capacity to grow, respond to stimuli, and reproduce. An entity with these properties is generally considered life. However, not every definition of life considers all of these properties to be essential. Human-made analogs of life may also be considered to be life.

The biosphere is the part of Earth's outer shell – including land, surface rocks, water, air and the atmosphere – within which life occurs, and which biotic processes in turn alter or transform. From the broadest geophysiological point of view, the biosphere is the global ecological system integrating all living beings and their relationships, including their interaction with the elements of the lithosphere (rocks), hydrosphere (water), and atmosphere (air). Currently the entire Earth contains over 75 billion tons (150 *trillion* pounds or about 6.8 x 10 kilograms) of biomass (life), which lives within various environments within the biosphere.

Over nine-tenths of the total biomass on Earth is plant life, on which animal life depends very heavily for its existence. More than 2 million species of plant and animal life have been identified to date, and estimates of the actual number of existing species range from several million to well over 50 million. The number of individual species of life is constantly in some degree of flux, with new species appearing and others ceasing to exist on a continual basis. The total number of species is presently in rapid decline.

Evolution

Life is only known to exist on the planet Earth. The origin of life is still a poorly understood process, but it is thought to have occurred about 3.9 to 3.5 billion years ago during the hadean or archean eons on a primordial earth that had a substantially different environment than is found at present. These life forms possessed the basic traits of self-replication and inheritable traits. Once life had appeared, the process of evolution by natural selection resulted in the development of ever-more diverse life forms.

Species that were unable to adapt to the changing environment and competition from other life forms became extinct. However, the fossil record retains evidence of many of these older species. Current fossil and DNA evidence shows that all existing species can trace a continual ancestry back to the first primitive life forms.

The advent of photosynthesis in very basic forms of plant life worldwide allowed the sun's energy to be harvested to create conditions allowing for more complex life. The resultant oxygen accumulated in the atmosphere and gave rise to the ozone layer. The incorporation of smaller cells within larger ones resulted in the development of yet more complex cells called eukaryotes. Cells within colonies became increasingly specialized, resulting in true multicellular organisms. With the ozone layer absorbing harmful ultraviolet radiation, life colonized the surface of Earth.

Microbes

The first form of life to develop on the Earth were microbes, and they remained the only form of life on the planet until about a billion years ago when multi-cellular organisms began to appear. Microorganisms are single-called organisms that are generally microscopic, and smaller than the human eye can see. They include Bacteria, Fungi, Archaea and Protista.

These life forms are found in almost every location on the Earth where there is liquid water, including the interior of rocks within the planet. Their reproduction is both rapid and profuse. The combination of a high mutation rate and a horizontal gene transfer ability makes them highly adaptable, and able to survive in new environments, including outer space. They form an essential part of the planetary ecosystem. However some microorganisms are pathogenic and can post health risk to other organisms.

Plants and Animals

Originally Aristotle divided all living things between plants, which generally do not move, and animals. In Linnaeus' system, these became the kingdoms Vegetabilia (later Plantae) and Animalia. Since then, it has become clear that the Plantae as originally defined included several unrelated groups, and the fungi and several groups of algae were removed to new kingdoms. However, these are still often considered plants in many contexts. Bacterial life is sometimes included in flora, and some classifications use the term *bacterial flora* separately from *plant flora.*

Among the many ways of classifying plants are by regional floras, which, depending on the purpose of study, can also include *fossil flora,* remnants of plant life from a previous era. People in many regions and countries take great pride in their individual arrays of characteristic flora, which can vary widely across the globe due to differences in climate and terrain.

Regional floras commonly are divided into categories such as *native flora* and *agricultural and garden flora,* the lastly mentioned of which are intentionally grown and cultivated. Some types of "native flora" actually have been introduced centuries ago by people migrating from one region or continent to another, and become an integral part of the native, or natural flora of the place to which they were introduced. This is an example of how human interaction with nature can blur the boundary of what is considered nature.

Another category of plant has historically been carved out for *weeds.* Though the term has fallen into disfavour among botanists as a formal way to categorize "useless" plants, the informal use of the word "weeds" to describe those plants that are deemed worthy of elimination is illustrative of the general tendency of people and societies to seek to alter or shape the course of nature. Similarly, animals are often categorized in ways such as *domestic, farm animals, wild animals, pests*, etc., according to their relationship to human life.

Animals as a category have several characteristics that generally set them apart from other living things, though not traced by scientists to having legs or wings instead of roots and leaves. Animals are eukaryotic and usually multicellular (although see Myxozoa), which separates them from bacteria, archaea and most protists. They are heterotrophic, generally digesting food in an internal chamber, which separates them from plants and algae. They are also distinguished from plants, algae, and fungi by lacking cell walls.

With a few exceptions, most notably the sponges (Phylum Porifera), animals have bodies differentiated into separate tissues. These include muscles, which are able to contract and control locomotion, and a nervous system, which sends and processes signals. There is also typically an internal digestive chamber. The eukaryotic cells possessed by all animals are surrounded by a characteristic extracellular matrix composed of collagen and elastic glycoproteins. This may be calcified to form structures like shells, bones, and spicules, a framework upon which cells can move about and be reorganized during development and maturation, and which supports the complex anatomy required for mobility.

Human Interrelationship

Although humans currently comprise only a minuscule proportion of the total living biomass on Earth, the human effect on nature is disproportionately large. Because of the extent of human influence, the boundaries between what humans regard as nature and "made environments" is not clear cut except at the extremes. Even at the extremes, the amount of natural environment that is free of discernible human influence is presently diminishing at an increasingly rapid pace.

The development of technology by the human race has allowed the greater exploitation of natural resources and has helped to alleviate some of the risk from natural hazards. In spite of this progress, however, the fate of human civilization remains closely linked to changes in the environment. There exists a highly complex feedback loop between the use of advanced technology and changes to the environment that are only slowly becoming understood. Man-made threats to the Earth's natural environment include pollution, deforestation, and disasters such as oil spills. Humans have contributed to the extinction of many plants and animals.

Humans employ nature for both leisure and economic activities. The acquisition of natural resources for industrial use remains the primary component of the world's economic system. Some activities, such as hunting and fishing, are used for both sustenance and leisure, often by different people. Agriculture was first adopted around the 9th millennium BCE. Ranging from food production to energy, nature influences economic wealth.

Although early humans gathered uncultivated plant materials for food and employed the medicinal properties of vegetation for healing, most modern human use of plants is through agriculture. The clearance of large tracts of land for crop growth has led to a significant reduction in the amount available of forestation and wetlands, resulting in the loss of habitat for many plant and animal species as well as increased erosion.

Aesthetics and Beauty

Beauty in nature has historically been a prevalent theme in art and books, filling large sections of libraries and bookstores. That nature has been depicted and celebrated by so much art, photography, poetry and other literature shows the strength with which many people associate nature and beauty. Reasons why this association exists, and what the association consists of, is studied by the branch of philosophy called aesthetics. Beyond certain basic characteristics that many philosophers agree about to explain what is seen as beautiful, the opinions are virtually endless. Nature and wildness have been important subjects in various epochs of world history. An early tradition of landscape art began in China during the Tang Dynasty (618-907). The tradition of representing nature *as it is* became one of the aims of Chinese painting and was a significant influence in Asian art.

Wilderness portrayals in art became more prevalent in the 1800s, especially in the works of the Romantic movement. British artists John Constable and JMW Turner turned their attention to capturing the beauty of the natural world in their paintings. Before that, paintings had been primarily of religious scenes or of human beings. William Wordsworth's poetry described the wonder of the natural world, which had formerly been viewed as a threatening place. Increasingly the valuing of nature became an aspect of Western culture. This artistic movement also coincided with the Transcendentalist movement in the Western world. A common classical idea of beautiful art involves the word mimesis, the imitation of nature. Also in the realm of ideas about beauty in nature is that the perfect is implied through symmetry, equal division, and other perfect mathematical forms and notions.

Matter and Energy

Some fields of science see nature as matter in motion, obeying certain laws of nature which science seeks to understand. For this reason the most fundamental science is generally understood to be "physics" – the name for which is still recognizable as meaning that it is the study of nature.

Matter is commonly defined as the substance of which physical objects are composed. It constitutes the observable universe. The visible components of the universe are now believed to compose only 4 percent of the total mass. The remainder is believed to consist of 23 percent cold dark matter and 73 percent dark energy. The exact nature of these components is still unknown and is currently under intensive investigation by physicists.

The behaviour of matter and energy throughout the observable universe appears to follow well-defined physical laws. These laws have been employed to produce cosmological models that successfully explain the structure and the evolution of the universe we can observe.

The mathematical expressions of the laws of physics employ a set of twenty physical constants that appear to be static across the observable universe. The values of these constants have been carefully measured, but the reason for their specific values remains a mystery.

Beyond Earth

Outer space, also simply called *space*, refers to the relatively empty regions of the universe outside the atmospheres of celestial bodies. *Outer* space is used to distinguish it from airspace (and terrestrial locations). There is no discrete boundary between the Earth's atmosphere and space, as the atmosphere gradually attenuates with increasing altitude. Outer space within the solar system is called interplanetary space, which passes over into interstellar space at what is known as the heliopause.

Outer space is certainly spacious, but it is far from empty. Outer space is sparsely filled with several dozen types of organic molecules discovered to date by microwave spectroscopy, blackbody radiation left over from the big bang and the origin of the universe, and cosmic rays, which include ionized atomic nuclei and various subatomic particles.

There is also some gas, plasma and dust, and small meteors. Additionally, there are signs of human life in outer space today, such as material left over from previous manned and unmanned launches which are a potential hazard to spacecraft. Some of this debris re-enters the atmosphere periodically.

Although the planet Earth is currently the only known body within the solar system to support life, current evidence suggests that in the distant past the planet Mars possessed bodies of liquid water on the surface. For a brief period in Mars' history, it may have also been capable of forming life. At present though, most of the water remaining on Mars is frozen. If life exists at all on Mars, it is most likely to be located underground where liquid water can still exist.

Conditions on the other terrestrial planets, Mercury and Venus, appear to be too harsh to support life as we know it. But it has been conjectured that Europa, the fourth-largest moon of Jupiter, may possess a sub-surface ocean of liquid water and could potentially host life.

Recently, the team of Stephane Udry have discovered a new planet named Gliese 581 d, which is an extrasolar planet orbiting the red dwarf star Gliese 581. Gliese 581 d appears to lie in the habitable zone of space surrounding the star, and therefore could possibly host life as we know it.

Geographical Distribution of Man: Varieties: Density of Population etc.

Population Density

Population density (in agriculture standing stock and standing crop) is a measurement of population per unit area or unit volume. It is frequently applied to living organisms, and particularly to humans. It is a key geographic term.

Biological Population Densities

Population density is population divided by total land area.

Low densities may cause an extinction vortex and lead to further reduced fertility. This is called the Allee effect after the scientist who identified it. Examples of the causes in low population densities include:

- Increased problems with locating mates
- Increased inbreeding.

Different species have different expected densities. R-selected species commonly have high population densities, while K-selected species may have lower densities. Low densities may be associated with specialized mate location adaptations such as specialized pollinators; as found in the orchid family (*Orchidaceae*).

Human Population Density

For humans, population density is the number of people per unit of area usually per square kilometre or mile (which may include or exclude cultivated or potentially productive area). Commonly this may be calculated for a county, city, country, another territory, or the entire world. The world's population is 6.8 billion, and Earth's total area (including land and water) is 510 million square kilometres (197 million square miles).

Therefore the worldwide human population density is 6.8 billion ÷ 510 million = 13.3 per km^2 (34.5 per sq. mile). If only the Earth's land area of 150 million km^2 (58 million sq. miles) is taken into account, then human population density increases to 45.3 per km^2 (117.2 per sq. mile). This calculation includes all continental and island land area, including Antarctica. If Antarctica is also excluded, then population density rises to 50 people per km^2 (129.28 per sq. mile). Considering that over half of the Earth's land mass consists of areas inhospitable to human inhabitation, such as deserts and high mountains, and that population tends to cluster around seaports and fresh water sources, this number by itself does not give any meaningful measurement of human population density.

Several of the most densely-populated territories in the world are city-states, microstates, micronations, or dependencies. These territories share a relatively small area and a high urbanization level, with an economically specialized city population drawing also on rural resources outside the area, illustrating the difference between high population density and overpopulation.

Cities with high population densities are, by some, considered to be overpopulated, though the extent to which this is the case depends on factors like quality of housing and infrastructure and access to resources. Most of the most densely-populated cities are in southern and eastern Asia, though Cairo and Lagos in Africa also fall into this category.

City population is, however, heavily dependent on the definition of "urban area" used: densities are often higher for the central municipality itself, than when more recently-developed and administratively unincorporated suburban communities are included, as in the concepts of agglomeration or metropolitan area, the latter including sometimes neighbouring cities. For instance, Milwaukee has a greater population density when just the inner city is measured, and not the surrounding suburbs as well.

Race

Race refers to the classification of humans into populations or ancestral groups based on various factors such as culture, social practice or heritable characteristics. Conceptions of race, as well as specific ways of grouping races, vary by culture and over time, effectively functioning as folk taxonomies. Race serves to indicate essential types of individuals or fuzzy sets of people's traits.

When used as a biological term, race describes genetically divergent populations of humans that can be marked by common phenotypic and genotypic traits. The notion of race is often used by forensic anthropologists, who analyse skeletal remains, and those involved in biomedical research and race-based medicine. The term *race* should not be confused with subspecies because *race* is not an official biological taxonomy category. All modern-day humans belong to the same subspecies of *Homo sapiens sapiens.*

There is no scientific basis for hierarchically organized categories based in race and ethnicity. Scholars in the fields of anthropology, sociology, biology, and genetics sometimes prefer to group shared traits along ethnic lines which correspond to a history of endogamy.

History

Definitions of race have varied across cultures and over time, and have been controversial for social, political and scientific reasons. Until the 19th century, race was thought by many to constitute an immutable and distinct type or species which shared particular racial characteristics, such as body constitution, temperament and mental capacities. These distinct races were popularly thought to constitute a hierarchical chain of life known as the Great Chain of Being, which was said to have been created by God, and which often saw the white, European races as closest to God in perfection.

The word "race," interpreted to mean common descent, was introduced into English in about 1580, from the Old French *rasse* (1512), from Italian *razza*, which may have been derived from the Arabic Word "ras" meaning the head of someone or something. In this context, "ras" points to the root or the head of selected species. The etymology can be further traced back to Latin *gens* or Arabic "gens" meaning (clan, stock, people) and *genus* (birth, descent, origin, race, stock, family) cognate with Greek genos "race, kind," and gonos "birth, offspring, stock [...]." This late origin for the English and French terms is consistent with the thesis that the concept of "race" as defining a very small number of groups of human beings based on lineage dates from the time of Christopher Columbus. Older concepts that were also at least partly based on common descent, such as nation and tribe, entail a much larger number of groupings.

The Age of Enlightenment saw a renewed focus on defining race as a biological concept. In the centuries that followed, attempts were made to specify racial types, and to determine racial origins and correlates. In reaction to the perceived excesses and abuses of such efforts, an international organization called UNESCO released in 1950 a statement rejecting previous theses that races were natural, discrete dividing lines in the human species. Most notably, it stated, "The biological fact of race and the myth of 'race' should be distinguished'. For all practical social purposes, 'race' is not so much a biological phenomenon as a social myth." As forensic anthropologists, biomedical researchers, and other scholars continued to revisit and refine the concept, some maintained that race denoted scientifically practical distinctions.

Modern Debates

Models of Human Evolution

In a 1995 article, Leonard Lieberman and Fatimah Jackson suggested that any new support for a biological concept of race will likely come from another source, namely, the study of human evolution. They therefore ask what, if any, implications current models of human evolution may have for any biological conception of race.

Today, all humans are classified as belonging to the species *Homo sapiens* and sub-species *Homo sapiens sapiens*. However, this is not the first species of hominids: the first species of genus *Homo*, Homo habilis, are theorized to have evolved in East Africa at least 2 million years ago, and members of this species populated different parts of Africa in a relatively short time. *Homo erectus* is theorized to have evolved more than 1.8 million years ago, and by 1.5 million years ago had spread throughout Europe and Asia. Virtually all physical anthropologists agree that *Homo sapiens* evolved out of *Homo erectus*.

Anthropologists have been divided as to whether *Homo sapiens* evolved as one interconnected species from *H. Erectus* (called the Multiregional Model, or the Regional Continuity Model), or evolved only in East Africa, and then migrated out of Africa and replaced *H. Erectus* populations throughout Europe and Asia (called the Out of Africa Model or the Complete Replacement Model). Anthropologists continue to debate both possibilities, and the evidence is technically ambiguous as to which model is correct, although most anthropologists currently favour the Out of Africa model.

Lieberman and Jackson argued that while advocates of both the Multiregional Model and the Out of Africa Model use the word race and make racial assumptions, none define the term. They conclude that "Each model has implications that both magnify and minimize the differences between races. Yet each model seems to take race and races as a conceptual reality. The net result is that those anthropologists who prefer to view races as a reality are encouraged to do so" and conclude that students of human evolution would be better off avoiding the word race, and instead describe genetic differences in terms of populations and clinal gradations.

Race as Subspecies

With the advent of the modern synthesis in the early 20th century, many biologists sought to use evolutionary models and populations genetics in an attempt to formalise taxonomy. The Biological Species Concept (BSC) is the most widely used system for describing species, this concept defines a species as a group of organisms that interbreed in their natural environment and produce viable offspring. In practice, species are not classified according to the BSC but according to typology by the use of a holotype, due to the difficulty of determining whether all members of a group of organisms do or can in practice potentially interbreed. BSC species are routinely classified on a subspecific level, though this classification is conducted differently for different taxons, for mammals the normal taxonomic unit below the species level is usually the subspecies.

More recently the Phylogenetic Species Concept (PSC) has gained a substantial following. The PSC is based on the idea of a least-inclusive taxonomic unit (LITU), in phylogenetic classification no subspecies can exist because they would automatically constitute a LITU (any monophyletic group). Technically species cease to exist as do all hierarchical taxa, a LITU is effectively defined as any monophyletic taxon, phylogenetics is strongly influenced by cladistics which classifies organisms based on evolution rather than similarities between groups of organisms. In biology the term "race" is used with caution because it can be ambiguous, "'Race' is not being defined or used consistently; its referents are varied and shift depending on context. The term is often used colloquially to refer to a range of human groupings. Religious, cultural, social, national, ethnic, linguistic, genetic, geographical and anatomical groups have been and sometimes still are called 'races'". Generally when it is used it is synonymous with subspecies. One main obstacle to identifying subspecies is that, while it is a recognised taxonomic term, it has no precise definition.

Species of organisms that are monotypic (i.e., form a single subspecies) display at least one of these properties:

- All members of the species are very similar and cannot be sensibly divided into biologically significant subcategories.
- The individuals vary considerably but the variation is essentially random and largely meaningless so far as genetic transmission of these variations is concerned (many plant species fit into this category, which is why horticulturists interested in preserving, say, a particular flower colour avoid propagation from seed, and instead use vegetative methods like propagation from cuttings).
- The variation among individuals is noticeable and follows a pattern, but there are no clear dividing lines among separate groups: they fade imperceptibly into one another. Such clinal variation displays a lack of allopatric partition between groups (i.e., a clearly defined boundary demarcating the subspecies), which is usually required before they are recognised as subspecies.

A *polytypic* species has two or more subspecies. These are separate populations that are more genetically different from one another and that are more reproductively isolated, gene flow between these populations is much reduced leading to genetic differentiation.

Morphological Subspecies

Traditionally subspecies are seen as geographically isolated and genetically differentiated populations. Or to put it another way "the designation 'subspecies' is used to indicate an objective degree of microevolutionary divergence" One objection to this idea is that it does not identify any degree of differentiation. Therefore, any population that is somewhat biologically different could be considered a subspecies, even to the level of a local population. As a result it is necessary to impose a threshold on the level of difference that is required for a population to be designated a subspecies.

This effectively means that populations of organisms must have reached a certain measurable level of difference to be recognised as subspecies. Dean Amadon proposed in 1949 that subspecies would be defined according to the seventy-five percent rule which means that 75% of a population must lie outside 99% of the range of other populations for a given defining morphological character or a set of characters. The seventy-five percent rule still has defenders but other scholars argue that it should be replaced with ninety or ninety-five percent rule.

In 1978, Sewall Wright suggested that human populations that have long inhabited separated parts of the world should, in general, be considered different subspecies by the usual criterion that most individuals of such populations can be allocated correctly by inspection. It does not require a trained anthropologist to classify an array of Englishmen, West Africans, and Chinese with 100% accuracy by features, skin colour, and type of hair despite so much variability within each of these groups that every individual can easily be distinguished from every other. However, it is customary to use the term race rather than subspecies for the major subdivisions of the human species as well as for minor ones.

On the other hand in practice subspecies are often defined by easily observable physical appearance, but there is not necessarily any evolutionary significance to these observed differences, so this form of classification has become less acceptable to evolutionary biologists. Likewise this typological approach to race is generally regarded as discredited by biologists and anthropologists.

Because of the difficulty in classifying subspecies morphologically, many biologists have found the concept problematic, citing issues such as:

- Visible physical differences do not always correlate with one another, leading to the possibility of different classifications for the same individual organisms.
- Parallel evolution can lead to the existence of the appearance of similarities between groups of organisms that are not part of the same species.
- Isolated populations within previously designated subspecies have been found to exist.
- The criteria for classification may be arbitrary if they ignore gradual variation in traits.

Subspecies Genetically Differentiated Populations

Another way to look at differences between populations is to measure genetic differences rather than physical differences. The Human Genome Project found only gradations in genetic variation, not sharp lines which would naturally define notions of race or ethnicity. "People who have lived in the same geographic region for many generations may have some alleles in common, but no allele will be found in all members of one population and in no members of any other."

Genetic differences between populations of organisms can be measured using the fixation index of Sewall Wright, which is often abbreviated to F_{ST}. This statistic is used to compare differences between any two given populations and can be used to measure genetic differences between populations for individual genes, or for many genes simultaneously. For example it is often stated that the fixation index for humans is about 0.15. This means that about 85% of the variation measured in the human population is within any population, and about 15% of the variation occurs between populations, or that any two individuals from different populations are almost as likely to be more similar to each other than either is to a member of their own group.

It is often stated that human genetic variation is low compared to other mammalian species, and it has been claimed that this should be taken as evidence that there is no natural subdivision of the human population. Wright himself believed that a value of 0.25 represented great genetic variation and that an F_{ST} of 0.15-0.25 represented moderate variation. It should, however, be noted that about 5% of human variation occurs between populations within continents, and therefore the F_{ST} between continental groups of humans (or races) is as low as 0.1 (or possibly lower).

In their 2003 paper "Human Genetic Diversity and the Nonexistence of Biological Races" Jeffrey Long and Rick Kittles give a long critique of the application of F_{ST} to human populations. They find that the figure of 85% is misleading because it implies that all human populations contain on average 85% of all genetic diversity. They claim that this does not correctly reflect human population history, because it treats all human groups as independent. A more realistic portrayal of the way human groups are related is to understand that some human groups are parental to other groups and that these groups represent paraphyletic groups to their descent groups. For example, under the recent African origin theory the human population in Africa is paraphyletic to all other human groups because it represents the ancestral group from which all non-African populations derive, but more than that, non-African groups only derive from a small non-representative sample of this African population.

This means that all non-African groups are more closely related to each other and to some African groups (probably east Africans) than they are to others, and further that the migration out of Africa represented a genetic bottleneck, with much of the diversity that existed in Africa not being carried out of Africa by the emigrating groups. This view produces a version of human population movements that do not result in all human populations being independent; but rather, produces a series of dilutions of diversity the further from Africa any population lives, each founding event representing a genetic subset of its parental population. Long and Kittles find that rather than 85% of human genetic diversity existing in all human populations, about 100% of human diversity exists in a single African population, whereas only about 70% of human genetic diversity exists in a population derived from New Guinea. Long and Kittles make the observation that this still produces a global human population that is genetically homogeneous compared to other mammalian populations.

Wright's F statistics are not used to determine whether a group can be described as a subspecies or not, though the statistic is used to measure the degree of differentiation between populations, the degree of genetic differentiation is not a marker of subspecies status. Generally taxonomists prefer to use phylogenetic analysis to determine whether a population can be considered a subspecies.

Phylogenetic analysis relies on the concept of derived characteristics that are not shared between groups, usually applying to populations that are allopatric (geographically separated) and therefore discretely bounded. This would make a subspecies, evolutionarily speaking, a clade - a group with a common evolutionary ancestor population. The smooth gradation of human genetic variation in general rules out any idea that human population groups can be considered monophyletic (cleanly divided) as there appears to always have been considerable gene flow between human populations.

Subspecies as clade;

By the 1970s many evolutionary scientists were avoiding the concept of "subspecies" as a taxonomic category for four reasons:

- Very few data indicate that contiguous subspecies ever become species
- Geographically disjunct groups regarded as subspecies usually can be demonstrated to actually be distinct species
- Subspecies had been recognized on the basis of only 2-5 phenotypic characters, which often were adaptations to local environments but which did not reflect the evolutionary differentiation of populations as a whole
- With the advent of molecular techniques used to get a better handle on genetic introgression (gene flow), the picture afforded by looking at genetic variation was often at odds with the phenotypic variation (as is the case with looking at genes versus percentage of epidermal melanin in human populations).

These criticisms have coincided with the rise of cladistics.

A clade is a taxonomic group of organisms consisting of a single common ancestor and all the descendants of that ancestor. Every creature produced by sexual reproduction has two immediate lineages, one maternal and one paternal. Whereas Carolus Linnaeus established a taxonomy of living organisms based on anatomical similarities and differences, cladistics seeks to establish a taxonomy—the phylogenetic tree—based on genetic similarities and differences and tracing the process of acquisition of multiple characteristics by single organisms.

A phylogenetic tree like the one shown above is usually derived from DNA or protein sequences from populations. Often mitochondrial DNA or Y chromosome sequences are used to study ancient human migration paths. These single-locus sources of DNA do not recombine and are inherited from a single parent. Individuals from the various continental groups tend to be more similar to one another than to people from other continents, and tracing either mitochondrial DNA or non-recombinant Y-chromosome DNA explains how people in one place may be largely derived from people in some remote location. The tree is rooted in the common ancestor of chimpanzees and humans, which is believed to have originated in Africa. Horizontal distance corresponds to two things:

1. Genetic distance. Given below the diagram, the genetic difference between humans and chimpanzees is roughly 2%, or 20 times larger than the variation among modern humans.
2. Temporal remoteness of the most recent common ancestor. Rough estimates are given above the diagram, in millions of years. The mitochondrial most recent common ancestor of

modern humans lived roughly 200,000 years ago, latest common ancestors of humans and chimpanzees between four and seven million years ago.

Chimpanzees and humans belong to different genera, indicated in Blue. Formation of species and subspecies is also indicated, and the formation of "races" is indicated in the green rectangle to the right (note that only a very rough representation of human phylogeny is given, and the points made in the preceding section, insofar as they apply to an "African race", are understood here). Note that vertical distances are not meaningful in this representation.

Most evolutionary scientists have rejected the identification of races with clades for two reasons. First, as Rachel Caspari (2003) argued, clades are by definition monophyletic groups (a taxon that includes *all* descendants of a given ancestor) since no groups currently regarded as races are monophyletic, none of those groups can be clades.

For anthropologists Lieberman and Jackson (1995), however, there are more profound methodological and conceptual problems with using cladistics to support concepts of race. They emphasize that "the molecular and biochemical proponents of this model explicitly use racial categories *in their initial grouping of samples*". For example, the large and highly diverse macroethnic groups of East Indians, North Africans, and Europeans are presumptively grouped as Caucasians prior to the analysis of their DNA variation.

This limits and skews interpretations, obscures other lineage relationships, deemphasizes the impact of more immediate clinal environmental factors on genomic diversity, and can cloud our understanding of the true patterns of affinity. They argue that however significant the empirical research, these studies use the term race in conceptually imprecise and careless ways. They suggest that the authors of these studies find support for racial distinctions only because they began by assuming the validity of race.

For empirical reasons we prefer to place emphasis on clinal variation, which recognizes the existence of adaptive human hereditary variation and simultaneously stresses that such variation is not found in packages that can be labelled *races*.

Indeed, recent research reports evidence for smooth, clinal genetic variation even in regions previously considered racially homogeneous, with the apparent gaps turning out to be artifacts of sampling techniques (Serre & Paabo 2004). These scientists do not dispute the importance of cladistic research, only its retention of the word race, when reference to populations and clinal gradations are more than adequate to describe the results.

Population Genetics: Population and Cline

At the beginning of the 20th century, anthropologists questioned, and eventually abandoned, the claim that biologically distinct races are isomorphic with distinct linguistic, cultural, and social groups. Shortly thereafter, the rise of population genetics provided scientists with a new understanding of the sources of phenotypic variation. This new science has led many mainstream evolutionary scientists in anthropology and biology to question the very validity of race as a scientific concept describing an objectively real phenomenon. Those who came to reject the validity of the concept of race did so for four reasons: empirical, definitional, the availability of alternative concepts, and ethical (Lieberman and Byrne 1993).

The first to challenge the concept of race on empirical grounds were anthropologists Franz Boas, who demonstrated phenotypic plasticity due to environmental factors (Boas 1912), and Ashley Montagu (1941, 1942), who relied on evidence from genetics. Zoologists Edward O. Wilson and W. Brown then challenged the concept from the perspective of general animal systematics, and further rejected the claim that "races" were equivalent to "subspecies" (Wilson and Brown 1953).

Clines

One crucial innovation in reconceptualizing genotypic and phenotypic variation was anthropologist C. Loring Brace's observation that such variations, insofar as it is affected by natural selection, migration, or genetic drift, are distributed along geographic gradations or clines (Brace 1964). In part this is due to isolation by distance. This point called attention to a problem common to phenotype-based descriptions of races (for example, those based on hair texture and skin colour): they ignore a host of other similarities and differences (for example, blood type) that do not correlate highly with the markers for race. Thus, anthropologist Frank Livingstone's conclusion, that since clines cross racial boundaries, "there are no races, only clines" (Livingstone 1962: 279).

In a response to Livingston, Theodore Dobzhansky argued that when talking about race one must be attentive to how the term is being used: "I agree with Dr. Livingston that if races have to be 'discrete units,' then there are no races, and if 'race' is used as an 'explanation' of the human variability, rather than vice versa, then the explanation is invalid." He further argued that one could use the term race if one distinguished between "race differences" and "the race concept."

The former refers to any distinction in gene frequencies between populations; the latter is "a matter of judgment." He further observed that even when there is clinal variation, "Race differences are objectively ascertainable biological phenomena... but it does not follow that racially distinct populations must be given racial (or subspecific) labels." In short, Livingston and Dobzhansky agree that there are genetic differences among human beings; they also agree that the use of the race concept to classify people, and how the race concept is used, is a matter of social convention. They differ on whether the race concept remains a meaningful and useful social convention.

In 1964, biologists Paul Ehrlich and Holm pointed out cases where two or more clines are distributed discordantly—for example, melanin is distributed in a decreasing pattern from the equator north and south; frequencies for the haplotype for beta-S hemoglobin, on the other hand, radiate out of specific geographical points in Africa (Ehrlich and Holm 1964). As anthropologists Leonard Lieberman and Fatimah Linda Jackson observed, "Discordant patterns of heterogeneity falsify any description of a population as if it were genotypically or even phenotypically homogeneous" (Lieverman and Jackson 1995).

Patterns such as those seen in human physical and genetic variation as described above, have led to the consequence that the number and geographic location of any described races is highly dependent on the importance attributed to, and quantity of, the traits considered. For example, if only skin colour and a "two race" system of classification were used, then one might classify Indigenous Australians in the same race as Black people, and Caucasians in the same race as East

Asian people, but biologists and anthropologists would dispute that these classifications have any scientific validity. Scientists discovered a skin-lighting mutation that partially accounts for the appearance of Light skin in humans (people who migrated out of Africa northward into what is now Europe) which they estimate occurred 20,000 to 50,000 years ago. The East Asians owe their relatively light skin to different mutations. On the other hand, the greater the number of traits (or alleles) considered, the more subdivisions of humanity are detected, since traits and gene frequencies do not always correspond to the same geographical location. Or as Ossario and Duster (2005) put it:

Anthropologists long ago discovered that humans' physical traits vary gradually, with groups that are close geographic neighbours being more similar than groups that are geographically separated. This pattern of variation, known as clinal variation, is also observed for many alleles that vary from one human group to another. Another observation is that traits or alleles that vary from one group to another do not vary at the same rate. This pattern is referred to as nonconcordant variation.

Because the variation of physical traits is clinal and nonconcordant, anthropologists of the late 19th and early 20th centuries discovered that the more traits and the more human groups they measured, the fewer discrete differences they observed among races and the more categories they had to create to classify human beings. The number of races observed expanded to the 30s and 50s, and eventually anthropologists concluded that there were no discrete races (Marks, 2002). Twentieth and 21st century biomedical researchers have discovered this same feature when evaluating human variation at the level of alleles and allele frequencies. Nature has not created four or five distinct, nonoverlapping genetic groups of people.

More recent genetic studies indicate that skin colour may change radically over as few as 100 generations, or about 2,500 years, given the influence of the environment.

Populations

Population geneticists have debated as to whether the concept of population can provide a basis for a new conception of race. In order to do this, a working definition of population must be found. Surprisingly, there is no generally accepted concept of population that biologists use. It has been pointed out that the concept of population is central to ecology, evolutionary biology and conservation biology, but also that most definitions of population rely on qualitative descriptions such as "a group of organisms of the same species occupying a particular space at a particular time" Waples and Gaggiotti identify two broad types of definitions for populations; those that fall into an *ecological paradigm*, and those that fall into an *evolutionary paradigm*. Examples of such definitions are:

- Ecological paradigm: A group of individuals of the same species that co-occur in space and time and have an opportunity to interact with each other.
- Evolutionary paradigm: A group of individuals of the same species living in close-enough proximity that any member of the group can potentially mate with any other member.

Richard Lewontin, claiming that 85 percent of human variation occurs within populations and not among populations, argued that neither "race" nor "subspecies" were appropriate or useful ways to describe populations (Lewontin 1973). Nevertheless, barriers—which may be cultural or physical— between populations can limit gene flow and increase genetic differences. Recent work by population geneticists conducting research in Europe suggests that ethnic identity can be a barrier to gene flow. Others, such as Ernst Mayr, have argued for a notion of "geographic race". Some researchers report the variation between racial groups (measured by Sewall Wright's population structure statistic F_{ST}) accounts for as little as 5% of human genetic variation[2]. Sewall Wright himself commented that if differences this large were seen in another species, they would be called subspecies. In 2003 A. W. F. Edwards argued that cluster analysis supersedes Lewontin's arguments.

These empirical challenges to the concept of race forced evolutionary sciences to reconsider their definition of race. Mid-century, anthropologist William Boyd defined race as: A population which differs significantly from other populations in regard to the frequency of one or more of the genes it possesses. It is an arbitrary matter which, and how many, gene loci we choose to consider as a significant "constellation" (Boyd 1950).

Lieberman and Jackson (1994) have pointed out that "the weakness of this statement is that if one gene can distinguish races then the number of races is as numerous as the number of human couples reproducing." Moreover, anthropologist Stephen Molnar has suggested that the discordance of clines inevitably results in a multiplication of races that renders the concept itself useless (Molnar 1992).

The distribution of many physical traits resembles the distribution of genetic variation within and between human populations (American Association of Physical Anthropologists 1996; Keita and Kittles 1997). For example, ~90% of the variation in human head shapes occurs within every human group, and ~10% separates groups, with a greater variability of head shape among individuals with recent African ancestors (Relethford 2002).

Conversely, in the paper "Genetic similarities within and between human populations" Witherspoon *et al.* (2007) show that even when individuals can be reliably assigned to specific population groups, it is still possible for two randomly chosen individuals from different populations/clusters to be more similar to each other than to a randomly chosen member of their own cluster. This is because multi locus clustering relies on population level similarities, rather than individual similarities, so that each individual is classified according to their similarity to the typical genotype for any given population. The paper claims that this masks a great deal of genetic similarity between individuals belonging to different clusters. Or in other words, two individuals from different clusters can be more similar to each other than to a member of their own cluster, while still both being more similar to the typical genotype of their own cluster than to the typical genotype of a different cluster.

When differences between individual pairs of people are tested, Witherspoon *et al.* found that the answer to the question "How often is a pair of individuals from one population genetically more dissimilar than two individuals chosen from two different populations?" is not adequately addressed by multi locus clustering analyses. They found that even for just three population groups separated by large geographic ranges (European, African and East Asian) the inclusion of many thousands of loci is required before the answer can become "never". On the other hand, the accurate classification of the global population must include more closely related and admixed populations, which will increase this above zero, so they state "In a similar vein, Romualdi et al. (2002) and Serre and Paabo (2004) have suggested that highly accurate classification of individuals from continuously sampled (and therefore closely related) populations may be impossible". Witherspoon *et al.* conclude "The fact that, given enough genetic data, individuals can be correctly assigned to their populations of origin is compatible with the observation that most human genetic variation is found within populations, not between them. It is also compatible with our finding that, even when the most distinct populations are considered and hundreds of loci are used, individuals are frequently more similar to members of other populations than to members of their own population"

Molecular Genetics: Lineages and Clusters

With the recent availability of large amounts of human genetic data from many geographically distant human groups, scientists have again started to investigate the relationships between people from various parts of the world. One method is to investigate DNA molecules that are passed down from mother to child (mtDNA) or from father to son (Y chromosomes). These form molecular lineages and can be informative regarding prehistoric population migrations. Alternatively, autosomal alleles are investigated in an attempt to understand how much genetic material groups of people share.

This work has led to a debate amongst geneticists, molecular anthropologists and medical doctors as to the validity of concepts such as race. Some researchers insist that classifying people into groups based on ancestry may be important from medical and social policy points of view, and claim to be able to do so accurately. Others claim that individuals from different groups share far too much of their genetic material for group membership to have any medical implications. This has reignited the scientific debate over the validity of human classification and concepts of race.

Current Views Across Disciplines

One result of debates over the meaning and validity of the concept of race is that the current literature across different disciplines regarding human variation lacks consensus, though within some fields, such as biology, there is strong consensus. Some studies use the word race in its early essentialist taxonomic sense. Many others still use the term race, but use it to mean a population, clade, or haplogroup. Others eschew the concept of race altogether, and use the concept of population as a less problematic unit of analysis.

Since 1932, an increasing number of college textbooks introducing physical anthropology have rejected race as a valid concept: from 1932 to 1976, only seven out of thirty-two rejected race; from 1975 to 1984, thirteen out of thirty-three rejected race; from 1985 to 1993, thirteen out of nineteen rejected race. According to one academic journal entry, where 78 percent of the articles in the 1931 *Journal of Physical Anthropology* employed these or nearly synonymous terms reflecting a bio-race paradigm, only 36 percent did so in 1965, and just 28 percent did in 1996. The *Statement on "Race"* (1998) composed by a select committee of anthropologists and issued by the executive board of the American Anthropological Association as a statement they "believe [...] represents generally the contemporary thinking and scholarly positions of a majority of anthropologists", declares:

With the vast expansion of scientific knowledge in this century, it has become clear that human populations are not unambiguous, clearly demarcated, biologically distinct groups. [...] Given what we know about the capacity of normal humans to achieve and function within any culture, we conclude that present-day inequalities between so-called "racial" groups are not consequences of their biological inheritance but products of historical and contemporary social, economic, educational, and political circumstances.

In an ongoing debate, some geneticists argue that race is neither a meaningful concept nor a useful heuristic device, and even that genetic differences among groups are biologically meaningless, because more genetic variation exists within such races than among them, and that racial traits overlap without discrete boundaries. Other geneticists, in contrast, argue that categories of self-identified race/ethnicity or biogeographic ancestry are both valid and useful, that these categories correspond to clusters inferred from multilocus genetic data, and that this correspondence implies that genetic factors might contribute to unexplained phenotypic variation between groups.

In February, 2001, the editors of the medical journal *Archives of Pediatrics and Adolescent Medicine* asked authors to no longer use race as an explanatory variable and not to use obsolescent terms. Some other peer-reviewed journals, such as the *New England Journal of Medicine* and the *American Journal of Public Health*, have made similar endeavours. Furthermore, the National Institutes of Health recently issued a program announcement for grant applications through February 1, 2006, specifically seeking researchers who can investigate and publicize among primary care physicians the detrimental effects on the nation's health of the practice of medical racial profiling using such terms. The program announcement quoted the editors of one journal as saying that, "analysis by race and ethnicity has become an analytical knee-jerk reflex." A survey, taken in 1985 (Lieberman *et al.* 1992), asked 1,200 American scientists how many disagree with the following proposition: "There are biological races in the species *Homo sapiens*." The responses were:

- biologists 16%
- developmental psychologists 36%
- physical anthropologists 41%
- cultural anthropologists 53%.

The figure for physical anthropologists at PhD granting departments was slightly higher, rising from 41% to 42%, with 50% agreeing. This survey, however, did not specify any particular definition of race (although it did clearly specify *biological race* within the *species Homo sapiens*); it is difficult to say whether those who supported the statement thought of race in taxonomic or population terms.

The same survey, taken in 1999, showed the following changing results for anthropologists:

- physical anthropologists 69%
- cultural anthropologists 80%.

On May 17, 1998 The American Anthropological Association produced a "Statement on 'Race'":

In the United States both scholars and the general public have been conditioned to viewing human races as natural and separate divisions within the human species based on visible physical differences. With the vast expansion of scientific knowledge in this century, however, it has become clear that human populations are not unambiguous, clearly demarcated, biologically distinct groups. Evidence from the analysis of genetics (e.g., DNA) indicates that most physical variation, about 94%, lies within so-called racial groups. Conventional geographic "racial" groupings differ from one another only in about 6% of their genes. This means that there is greater variation within "racial" groups than between them. In neighbouring populations there is much overlapping of genes and their phenotypic (physical) expressions. Throughout history whenever different groups have come into contact, they have interbred. The continued sharing of genetic materials has maintained all of humankind as a single species.

In Poland the race concept was rejected by only 25 percent of anthropologists in 2001, although: "Unlike the U.S. anthropologists, Polish anthropologists tend to regard race as a term without taxonomic value, often as a substitute for population."

In the face of these issues, some evolutionary scientists have simply abandoned the concept of race in favour of "population." What distinguishes population from previous groupings of humans by race is that it refers to a breeding population (essential to genetic calculations) and not to a biological taxon. Other evolutionary scientists have abandoned the concept of race in favour of cline (meaning, how the frequency of a trait changes along a geographic gradient). (The concepts of population and cline are not, however, mutually exclusive and both are used by many evolutionary scientists.)

According to Jonathan Marks, By the 1970s, it had become clear that (1) most human differences were cultural; (2) what was not cultural was principally polymorphic - that is to say, found in diverse groups of people at different frequencies; (3) what was not cultural or polymorphic was principally clinal- that is to say, gradually variable over geography; and (4) what was left - the component of human diversity that was not cultural, polymorphic, or clinal - was very small.

A consensus consequently developed among anthropologists and geneticists that race as the previous generation had known it - as largely discrete, geographically distinct, gene pools - did not exist.

In the face of this rejection of race by evolutionary scientists, many social scientists have replaced the word race with the word "ethnicity" to refer to self-identifying groups based on beliefs concerning shared culture, ancestry and history. Alongside empirical and conceptual problems with "race," following the Second World War, evolutionary and social scientists were acutely aware of how beliefs about race had been used to justify discrimination, apartheid, slavery, and genocide. This questioning gained momentum in the 1960s during the U.S. Civil rights movement and the emergence of numerous anti-colonial movements worldwide. They thus came to believe that race itself is a social construct, a concept that was believed to correspond to an objective reality but which was believed in because of its social functions.

Races as Social Constructions

Even as the idea of race was becoming a powerful organizing principle in many societies, some observers criticized the concept. In Europe, the gradual transition in appearances from one group to adjacent groups suggested to Blumenbach that "one variety of mankind does so sensibly pass into the other, that you cannot mark out the limits between them" (Marks 1995). As anthropologists and other evolutionary scientists have shifted away from the language of race to the term *population* to talk about genetic differences, Historians, anthropologists and social scientists have re-conceptualized the term "race" as a cultural category or social construct, in other words, as a particular way that some people have of talking about themselves and others.

Dr. Craig Venter and scientist Francis Collins of the National Institute of Health jointly made the announcement of the mapping of the human genome in 2000. Upon examining the data from the genome mapping, he realized that although we are indeed further apart genetically from each other, (1-3% instead of the assumed 1%), the types of variations don't warrant calling each other different races. Venter says quote..” Race is a social concept. It’s not a scientific one. There are no bright lines (that would stand out), if we could compare all the sequenced genomes of everyone on the planet.” “When we try to apply science to try to sort out these social differences, it all falls apart.”

Stephan Palmie has recently summarized, race “is not a thing but a social relation”; or, in the words of Katya Gibel Mevorach, “a metonym,” “a human invention whose criteria for differentiation are neither universal nor fixed but have always been used to manage difference.” As such it cannot be a useful analytical concept; rather, the use of the term “race” itself must be analyzed. Moreover, they argue that biology will not explain why or how people use the idea of race: history and social relationships will.

In the United States

The immigrants to the Americas came ultimately from every region of Europe, Africa, and Asia. Throughout America the immigrants mixed among themselves and with the indigenous inhabitants of the continent. In the United States, for example, most people who self-identify as African American have some European ancestors—in one analysis of genetic markers that have differing frequencies between continents, European ancestry ranged from an estimated 7% for a sample of Jamaicans to <“23% for a sample of African Americans from New Orleans (Parra *et al.* 1998). Similarly, many people who identify as European American have some African or Native American ancestors, either through openly interracial marriages or through the gradual inclusion of people with mixed ancestry into the majority population. In a survey of college students who self-identified as white in a northeastern U.S. University, <“30% were estimated to have less than 90% European ancestry.

Since the early history of the United States, Native Americans, African Americans, and European Americans have been classified as belonging to different races. For nearly three centuries, the criteria for membership in these groups were similar, comprising a person’s appearance, his fraction of known non-White ancestry, and his social circle. But the criteria for membership in these races diverged in the late 19th century. During Reconstruction, increasing numbers of Americans began to consider anyone with “one drop” of known “Black blood” to be Black, regardless of appearance. By the early 20th century, this notion of invisible blackness was made statutory in many states and widely adopted nationwide. In contrast, Amerindians continue to be defined by a certain percentage of "Indian blood" (called *blood quantum*), due in large part to American slavery ethics. Finally, to be White one had to have perceived "pure" White ancestry.

Efforts to sort the increasingly mixed population of the United States into discrete categories generated many difficulties (Spickard 1992). Efforts to track mixing between groups led to a proliferation of categories, such as mulatto and octoroon, and blood quantum distinctions that became increasingly untethered from self-reported ancestry. A person's racial identity can change over time, and self-ascribed race can differ from assigned race (Kressin *et al.* 2003).

The difference between how Native American and Black identities are defined today (blood quantum versus one-drop rule) has demanded explanation. According to anthropologists such as Gerald Sider, the goal of such racial designations was to concentrate power, wealth, privilege and land in the hands of Whites in a society of White hegemony and privilege. The differences have little to do with biology and far more to do with the history of racism and specific forms of White supremacy (the social, geopolitical and economic agendas of dominant Whites vis-a-vis subordinate Blacks and Native Americans), especially the different roles Blacks and Amerindians occupied in White-dominated 19th century America.

The theory suggests that the blood quantum definition of Native American identity enabled Whites to acquire Amerindian lands, while the one-drop rule of Black identity enabled Whites to preserve their agricultural labour force. The contrast presumably emerged because, as peoples transported far from their land and kinship ties on another continent, Black labour was relatively easy to control, thus reducing Blacks to valuable commodities as agricultural labourers. In contrast, Amerindian labour was more difficult to control; moreover, Amerindians occupied large territories that became valuable as agricultural lands, especially with the invention of new technologies such as railroads; thus, the blood quantum definition enhanced White acquisition of Amerindian lands in a doctrine of Manifest Destiny that subjected them to marginalization and multiple episodic localized campaigns of extermination.

The political economy of race had different consequences for the descendants of aboriginal Americans and African slaves. The 19th century blood quantum rule meant that it was relatively easier for a person of mixed Euro-Amerindian ancestry to be accepted as White. The offspring of only a few generations of intermarriage between Amerindians and Whites likely would not have been considered

Amerindian at all (at least not in a legal sense). Amerindians could have treaty rights to land, but because an individual with one Amerindian great-grandparent no longer was classified as Amerindian, they lost any legal claim to Amerindian land. According to the theory, this enabled Whites to acquire Amerindian lands. The irony is that the same individuals who could be denied legal standing because they were "too White" to claim property rights, might still be Amerindian enough to be considered "breeds", stigmatized for their Native American ancestry.

The one-drop rule, on the other hand, made it relatively difficult for anyone of known Black ancestry to be accepted as White during the 20th century. The child of a Black sharecropper and a White person was considered Black. And, significantly, in terms of the economics of sharecropping, such a person also would likely be a sharecropper as well, thus adding to the employer's labour force. However, some people with hints of African ancestry had light enough skin and other features to pass as "white". This did not happen for many people though. Many African-Americans today still possess small traces of European ancestry because of this mixing of races.

In short, this theory suggests that in a 20th century economy that benefited from sharecropping, it was useful to have as many Blacks as possible. Conversely, in a 19th century nation bent on westward expansion, it was advantageous to diminish the numbers of those who could claim title to Amerindian lands by simply defining them out of existence.

It must be mentioned, however, that although some scholars of the Jim Crow period agree that the 20th century notion of invisible Blackness shifted the colour line in the direction of paleness, thereby swelling the labour force in response to Southern Blacks' Great Migration northwards, others (Joel Williamson, C. Vann Woodward, George M. Fredrickson, Stetson Kennedy) see the one-drop rule as a simple consequence of the need to define Whiteness as being pure, thus justifying White-on-Black oppression.

In any event, over the centuries when Whites wielded power over both Blacks and Amerindians and widely believed in their inherent superiority over people of colour, it is no coincidence that the hardest racial group in which to prove membership was the White one.

In the United States, social and legal conventions developed over time that forced individuals of mixed ancestry into simplified racial categories (Gossett 1997). An example is the aforementioned one-drop rule implemented in some state laws that treated anyone with a single known African American ancestor as black (Davis 2001).

The decennial censuses conducted since 1790 in the United States also created an incentive to establish racial categories and fit people into those categories (Nobles 2000). In other countries in the Americas where mixing among groups was overtly more extensive, social categories have tended to be more numerous and fluid, with people moving into or out of categories on the basis of a combination of socioeconomic status, social class, ancestry, and appearance (Morner 1967).

The term "Hispanic" as an ethnonym emerged in the 20th century with the rise of migration of labourers from American Spanish-speaking countries to the United States. Today, the word "Latino" is often used as a synonym for "Hispanic". The definitions of both terms are non-race specific, and include people who consider themselves to be of distinct races (Black, White, Amerindian, Asian, and mixed groups). In contrast to "Latino" or "Hispanic", "Anglo" refers to non-Hispanic White Americans or non-Hispanic European Americans, most of whom speak the English language but are not necessarily of English descent.

In Brazil

Compared to 19th century United States, 20th century Brazil was characterized by a perceived relative absence of sharply defined racial groups. According to anthropologist Marvin Harris (1989), this pattern reflects a different history and different social relations. Basically, race in Brazil was "biologized," but in a way that recognized the difference between ancestry (which determines genotype) and phenotypic differences. There, racial identity was not governed by rigid descent rule, such as the one-drop rule, as it was in the United States. A Brazilian child was never automatically identified with the racial type of one or both parents, nor were there only a very limited number of categories to choose from.

Over a dozen racial categories would be recognized in conformity with all the possible combinations of hair colour, hair texture, eye colour, and skin colour. These types grade into each other like the colours of the spectrum, and no one category stands significantly isolated from the rest. That is, race referred preferentially to appearance, not heredity. The complexity of racial classifications in Brazil reflects the extent of miscegenation in Brazilian society, a society that remains highly, but not strictly, stratified along colour lines. Henceforth, the Brazilian narrative of a perfect "post-racist" country, must be met with caution, as sociologist Gilberto Freyre demonstrated in 1933 in *Casa Grande e Senzala*.

Marketing of Race: Genetic Lineages as Social Lineages

New research in molecular genetics, and the marketing of genetic identities through the analysis of one's Y chromosome, mtDNA or autosomal DNA, has reignited the debate surrounding race. Most of the controversy surrounds the question of how to interpret these new data, and whether conclusions based on existing data are sound. Although the vast majority of researchers endorse the view that continental groups do not constitute different subspecies, and molecular geneticists generally reject the identification of mtDNA and Y chromosomal lineages or allele clusters with "races", some anthropologists have suggested that the marketing of genetic analysis to the general public in the form of "Personalized Genetic Histories" (PGH) is leading to a new social construction of race.

Typically, a consumer of a commercial PGH service sends in a sample of DNA which is analyzed by molecular biologists and is sent a report, of which the following is a sample

"African DNA Ancestry Report"

The subject's likely haplogroup L2 is associated with the so-called Bantu expansion from West and Central sub-Saharan Africa east and south, dated 2,000-4,000 years ago... Between the 15th and 19th centuries C.E, the Atlantic slave trade resulted in the forced movement of approximately 13 million people from Africa, mainly to the Americas. Only approximately 11 million survived the passage and many more died in the early years of captivity. Many of these slaves were traded to the West African Cape Verde ports of embarkation through Portuguese and Arab middlemen and came from as far south as Angola. Among the African tribal groups, all Bantu-speaking, in which L2 is common are: Hausa, Kanuri, Fulfe, Songhai, Malunjin (Angola), Yoruba, Senegalese, Serer and Wolof.

Although no single sentence in such a report is technically wrong, through the combination of these sentences, anthropologists and others have argued, the report is telling a story that connects a haplotype with a language and a group of tribes. This story is generally rejected by research scientists because an individual receives his or her Y chromosome or mtDNA from only one ancestor in every generation; consequently, with every generation one goes back in time, the percentage of one's ancestors it represents halves; if one goes back hundreds (let alone thousands) of years, it represents only a tiny fragment of one's ancestry. As Mark Shriver and Rick Kittles recently remarked,

For many customers of lineage-based tests, there is a lack of understanding that their maternal and paternal lineages do not necessarily represent their entire genetic make-up. For example, an individual might have more than 85% Western European 'genomic' ancestry but still have a West African mtDNA or NRY lineage.

Nevertheless, they acknowledge, such stories are increasingly appealing to the general public. Thus, in his book *Blood of the Isles* (published in the US and Canada as *Saxons, Vikings and Celts: The Genetic Roots of Britain and Ireland*), however, Bryan Sykes discusses how people who have been mtDNA tested by his commercial laboratory and been found to belong to the same haplogroup have parties together because they see this as some sort of "bond", even though these people may not actually share very much ancestry.

Through these kinds of reports, new advances in molecular genetics are being used to create or confirm stories have about social identities. Although these identities are not racial in the biological sense, they are in the cultural sense in that they link biological and cultural identities. Nadia Abu el-Haj has argued that the significance of genetic lineages in popular conceptions of race owes to the perception that while genetic lineages, like older notions of race, suggests some idea of biological relatedness, unlike older notions of race they are not directly connected to claims about human behaviour or character. Abu el-Haj has thus argued that "postgenomics does seem to be giving race a new lease on life." Nevertheless, Abu el-Haj argues that to understand what it means to think of race in terms of genetic lineages or clusters, one must understand that

Race science was never just about classification. It presupposed a distinctive relationship between "nature" and "culture," understanding the differences in the former to ground and to generate the different kinds of persons ("natural kinds") and the distinctive stages of cultures and civilizations that inhabit the world.

Abu el-Haj argues that genomics and the mapping of lineages and clusters liberates "the new racial science from the older one by disentangling ancestry from culture and capacity." As an example, she refers to recent work by Hammer *et al.*, which aimed to test the claim that present-day Jews are more closely related to one another than to neighbouring non-Jewish populations. Hammer *et al.* found that the degree of genetic similarity among Jews shifted depending on the locus investigated, and suggested that this was the result of natural selection acting on particular loci. They therefore focused on the non-recombining Y chromosome to "circumvent some of the complications associated with selection".

As another example she points to work by Thomas *et al.*, who sought to distinguish between the Y chromosomes of Jewish priests (Kohanim), (in Judaism, membership in the priesthood is passed on through the father's line) and the Y chromosomes of non-Jews. Abu el-Haj concluded that this new "race science" calls attention to the importance of "ancestry" (narrowly defined, as it does not include all ancestors) in some religions and in popular culture, and people's desire to use science to confirm their claims about ancestry; this "race science," she argues, is fundamentally different from older notions of race that were used to explain differences in human behaviour or social status:

As neutral markers, junk DNA cannot generate cultural, behavioural, or, for that matter, truly biological differences between groups... mtDNA and Y-chromosome markers relied on in such work are not "traits" or "qualities" in the old racial sense. They do not render some populations more prone to violence, more likely to suffer psychiatric disorders, or for that matter, incapable of being fully integrated - because of their lower evolutionary development - into a European cultural world. Instead, they are "marks," signs of religious beliefs and practices... it is via biological noncoding genetic evidence that one can demonstrate that history itself is shared, that historical traditions are (or might well be) true."

On the other hand, there are tests that do not rely on molecular lineages, but rather on correlations between allele frequencies, often when allele frequencies correlate these are called clusters. Clustering analyses are less powerful than lineages because they cannot tell a historical story, they can only estimate the proportion of a person's ancestry from any given large geographical region. These sorts of tests use informative alleles called Ancestry-informative marker (AIM), which although shared across all human populations vary a great deal in frequency between groups of people living in geographically distant parts of the world.

These tests use contemporary people sampled from certain parts of the world as references to determine the likely proportion of ancestry for any given individual. In a recent Public Service Broadcasting (PBS) programme on the subject of genetic ancestry testing the academic Henry Louis Gates: "wasn't thrilled with the results (it turns out that 50 percent of his ancestors are likely European)". Charles Rotimi, of Howard University's National Human Genome Centre, is one of many who have highlighted the methodological flaws in such research—that "the nature or appearance of genetic clustering (grouping) of people is a function of how populations are sampled, of how criteria for boundaries between clusters are set, and of the level of resolution used" all bias the results—and concluded that people should be very cautious about relating genetic lineages or clusters to their own sense of identity.

Thus, in analyses that assign individuals to groups it becomes less apparent that self-described racial groups are reliable indicators of ancestry. One cause of the reduced power of the assignment of individuals to groups is admixture. For example, self-described African Americans tend to have a mix of West African and European ancestry. Shriver *et al.* (2003) found that on average African Americans have ~80% African ancestry. Also, in a survey of college students who self-identified as "white" in a northeastern U.S. university, ~30% of whites had less than 90% European ancestry.

Stephan Palmie has responded to Abu el-Haj's claim that genetic lineages make possible a new, politically, economically, and socially benign notion of race and racial difference by suggesting that efforts to link genetic history and personal identity will inevitably "ground present social arrangements in a time-hallowed past," that is, use biology to explain cultural differences and social inequalities.

Race and Intelligence

Researchers have reported differences in the average IQ test scores of various ethnic groups. The interpretation, causes, accuracy and reliability of these differences are highly controversial. Some researchers, such as Arthur Jensen, Richard Herrnstein, and Richard Lynn, have argued that such differences are at least partially genetic. Others, for example Thomas Sowell, argue that the differences largely owe to social and economic inequalities. Still others such as Stephen Jay Gould and Richard Lewontin have argued that categories such as "race" and "intelligence" are cultural constructs that render any attempt to explain such differences (whether genetically or sociologically) meaningless.

Political and Practical Uses

In Biomedicine

There is an active debate among biomedical researchers about the meaning and importance of race in their research. The primary impetus for considering race in biomedical research is the possibility of improving the prevention and treatment of diseases by predicting hard-to-ascertain factors on the basis of more easily ascertained characteristics. Some have argued that without cheap and widespread genetic tests, racial identification is the best way to predict for certain diseases, such as Cystic fibrosis, Lactose intolerance, Tay-Sachs Disease and sickle cell anemia, which are genetically linked and more prevalent in some populations than others. The most well-known examples of genetically determined disorders that vary in incidence among populations would be sickle cell disease, thalassaemia, and Tay-Sachs disease.

There has been criticism of associating disorders with race. For example, in the United States sickle cell is typically associated with black people, but this trait is also found in people of Mediterranean, Middle Eastern or Indian ancestry. The sickle cell trait offers some resistance to malaria. In regions where malaria is present sickle cell has been positively selected and consequently the proportion of people with it is greater. Therefore, it has been argued that sickle cell should not be associated with a particular race, but with having ancestors who lived in a malaria-prone region. Africans living in areas where there is no malaria, such as the East African highlands, have a prevalence of sickle cell as low as parts of Northern Europe.

Another example of the use of race in medicine is the recent U.S. FDA approval of BiDil, a medication for congestive heart failure targeted at black people in the United States. Several researchers have questioned the scientific basis for arguing the merits of a medication based on race, however. As Stephan Palmie has recently pointed out, black Americans were disproportionately affected by Hurricane Katrina, but for social and not climatological reasons; similarly, certain diseases may disproportionately affect different races, but not for biological reasons. Several researchers have suggested that BiDil was re-designated as a medicine for a race-specific illness because its manufacturer, Nitromed, needed to propose a new use for an existing medication to justify an extension of its patent and thus monopoly on the medication, not for pharmacological reasons.

Gene flow and intermixture also have an effect on predicting a relationship between race and “race linked disorders”. Multiple sclerosis is typically associated with people of European descent and is of low risk to people of African descent. However, due to gene flow between the populations, African Americans have elevated levels of MS relative to Africans. Notable African Americans affected by MS include Richard Pryor and Montel Williams. As populations continue to mix, the role of socially constructed races may diminish in identifying diseases.

A problem with making this distinction between Africans and Americans of African descent, however are recent discoveries of a link of where someone grew up as a child based upon latitudinal distance from the equator and a link with developing MS as an adult. While the incidence of MS increases with degree of melanin in the skin and Vitamin D production from sunlight exposure, in this manner the importance of race exists primarily in regards to childhood development and sunlight exposure. Rates of MS among caucasians and other races are also higher among adults who grew up in higher latitudes. This would make expression of genotype more important than the genotype itself since all races experience greater rates of development of MS if childhood exposure to sunlight and Vitamin D production is decreased.

In Law Enforcement

In an attempt to provide general descriptions that may facilitate the job of law enforcement officers seeking to apprehend suspects, the United States FBI employs the term “race” to summarize the general appearance (skin colour, hair texture, eye shape, and other such easily noticed characteristics) of individuals whom they are attempting to apprehend. From the perspective of law enforcement officers, it is generally more important to arrive at a description that will readily suggest the general appearance of an individual than to make a scientifically valid categorization by DNA or other such means. Thus, in addition to assigning a wanted individual to a racial category, such a description will include: height, weight, eye colour, scars and other distinguishing characteristics.

British Police use a classification based in the ethnic background of British society: W1 (White-British), W2 (White-Irish), W9 (Any other white background); M1 (White and black Caribbean), M2 (White and black African), M3 (White and Asian), M9 (Any other mixed background); A1 (Asian-Indian), A2 (Asian-Pakistani), A3 (Asian-Bangladeshi), A9 (Any other Asian background); B1 (Black Caribbean), B2 (Black African), B3 (Any other black background); O1 (Chinese),

O9 (Any other). Some of the characteristics that constitute these groupings are biological and some are learned (cultural, linguistic, etc.) traits that are easy to notice.

In many countries, such as France, the state is legally banned from maintaining data based on race, which often makes the police issue wanted notices to the public that include labels like "dark skin complexion", etc..

In the United States, the practice of racial profiling has been ruled to be both unconstitutional and also to constitute a violation of civil rights. There is active debate regarding the cause of a marked correlation between the recorded crimes, punishments meted out, and the country's populations. Many consider *de facto* racial profiling an example of institutional racism in law enforcement. The history of misuse of racial categories to impact adversely one or more groups and/or to offer protection and advantage to another has a clear impact on debate of the legitimate use of known phenotypical or genotypical characteristics tied to the presumed race of both victims and perpetrators by the government.

More recent work in racial taxonomy based on DNA cluster analysis has led law enforcement to narrow their search for individuals based on a range of phenotypical characteristics found consistent with DNA evidence.

While controversial, DNA analysis has been successful in helping police identify both victims and perpetrators by indicating what phenotypical characteristics to look for and what community the individual may have lived in. For example, in one case phenotypical characteristics suggested that the friends and family of an unidentified victim would be found among the Asian community, but the DNA evidence directed official attention to missing Native Americans, where her true identity was eventually confirmed. In an attempt to avoid potentially misleading associations suggested by the word "race," this classification is called "biogeographical ancestry" (BGA), but the terms for the BGA categories are similar to those used as for race.

The difference is that ancestry-informative DNA markers identify continent-of-ancestry admixture, not ethnic self-identity, and provide a wide range of phenotypical characteristics such that some people in a biogeographical category will not match the stereotypical image of an individual belonging to the corresponding race.

To facilitate the work of officials trying to find individuals based on the evidence of their DNA traces, firms providing the genetic analyses also provide photographs showing a full range of phenotypical characteristics of people in each biogeographical group. Of special interest to officials trying to find individuals on the basis of DNA samples that indicate a diverse genetic background is what range of phenotypical characteristics people with that general mixture of genotypical characteristics may display.

Forensic Anthropology

Similarly, forensic anthropologists draw on highly heritable morphological features of human remains (e.g. cranial measurements) to aid in the identification of the body, including in terms of race. In a recent article anthropologist Norman Sauer asked, "if races don't exist, why are forensic anthropologists so good at identifying them?" Sauer observed that the use of 19th century racial categories is widespread among forensic anthropologists:

- "In many cases there is little doubt that an individual belonged to the Negro, Caucasian, or Mongoloid racial stock."
- "Thus the forensic anthropologist uses the term race in the very broad sense to differentiate what are commonly known as white, black and yellow racial stocks."
- "In estimating race forensically, we prefer to determine if the skeleton is Negroid, or Non-Negroid. If findings favour Non-Negroid, then further study is necessary to rule out Mongoloid."

According to Sauer, "The assessment of these categories is based upon copious amounts of research on the relationship between biological characteristics of the living and their skeletons." Nevertheless, he says he agrees with other anthropologists that race is not a valid biological taxonomic category, and that races are socially constructed. He argued there is nevertheless a strong relationship between the phenotypic features forensic anthropologists base their identifications on, and popular racial categories. Thus, he argued, forensic anthropologists apply a racial label to human remains because their analysis of physical morphology enables them to predict that when the person was alive, a particular racial label would have been applied to them.

7

Geographical Distribution of Animals

Animals are a major group of mostly multicellular, eukaryotic organisms of the kingdom Animalia or Metazoa. Their body plan eventually becomes fixed as they develop, although some undergo a process of metamorphosis later on in their life. Most animals are motile, meaning they can move spontaneously and independently. All animals are also heterotrophs, meaning they must ingest other organisms for sustenance.

Etymology

The word "animal" comes from the Latin word *animal* (meaning *with soul*, from *anima*, soul). In everyday colloquial usage, the word usually refers to non-human animals. Frequently only closer relatives of humans such as mammals and other vertebrates are meant in colloquial use. The biological definition of the word refers to all members of the Kingdom Animalia, encompassing creatures ranging from insects to humans.

Characteristics

Animals have several characteristics that set them apart from other living things. Animals are eukaryotic and are mostly multicellular (although see Myxozoa), which separates them from bacteria and most protists. They are heterotrophic, generally digesting food in an internal chamber, which separates them from plants and algae. They are also distinguished from plants, algae, and fungi by lacking rigid cell walls. All animals are motile, if only at certain life stages. In most animals, embryos pass through a blastula stage, which is a characteristic exclusive to animals.

Structure

With a few exceptions, most notably the sponges (Phylum Porifera) and Placozoa, animals have bodies differentiated into separate tissues. These include muscles, which are able to contract and control locomotion, and nerve tissue, which sends and processes signals. There is also typically an internal digestive chamber, with one or two openings. Animals with this sort of organization are called metazoans, or eumetazoans when the former is used for animals in general.

All animals have eukaryotic cells, surrounded by a characteristic extracellular matrix composed of collagen and elastic glycoproteins. This may be calcified to form structures like shells, bones, and spicules. During development it forms a relatively flexible framework upon which cells can move about and be reorganized, making complex structures possible. In contrast, other multicellular organisms like plants and fungi have cells held in place by cell walls, and so develop by progressive growth. Also, unique to animal cells are the following intercellular junctions: tight junctions, gap junctions, and desmosomes.

Reproduction and Development

Nearly all animals undergo some form of sexual reproduction. They have a few specialized reproductive cells, which undergo meiosis to produce smaller motile spermatozoa or larger non-motile ova. These fuse to form zygotes, which develop into new individuals.

Many animals are also capable of asexual reproduction. This may take place through parthenogenesis, where fertile eggs are produced without mating, or in some cases through fragmentation.

A zygote initially develops into a hollow sphere, called a blastula, which undergoes rearrangement and differentiation. In sponges, blastula larvae swim to a new location and develop into a new sponge. In most other groups, the blastula undergoes more complicated rearrangement. It first invaginates to form a gastrula with a digestive chamber, and two separate germ layers — an external ectoderm and an internal endoderm. In most cases, a mesoderm also develops between them. These germ layers then differentiate to form tissues and organs.

Food and Energy Sourcing

All animals are heterotrophs, meaning that they feed directly or indirectly on other living things. They are often further subdivided into groups such as carnivores, herbivores, omnivores, and parasites.

Predation is a biological interaction where a predator (a heterotroph that is hunting) feeds on its prey (the organism that is attacked). Predators may or may not kill their prey prior to feeding on them, but the act of predation always results in the death of the prey. The other main category of consumption is detritivory, the consumption of dead organic matter. It can at times be difficult to separate the two feeding behaviours, for example where parasitic species prey on a host organism and then lay their eggs on it for their offspring to feed on its decaying corpse. Selective pressures imposed on one another has led to an evolutionary arms race between prey and predator, resulting in various antipredator adaptations.

Most animals feed indirectly from the energy of sunlight. Plants use this energy to convert sunlight into simple sugars using a process known as photosynthesis. Starting with the molecules carbon dioxide (CO_2) and water (H_2O), photosynthesis converts the energy of sunlight into chemical energy stored in the bonds of glucose ($C_6H_{12}O_6$) and releases oxygen (O_2). These sugars are then used as the building blocks which allow the plant to grow. When animals eat these plants (or eat other animals which have eaten plants), the sugars produced by the plant are used by the animal. They are either used directly to help the animal grow, or broken down, releasing stored solar energy, and giving the animal the energy required for motion. This process is known as glycolysis.

Animals living close to hydrothermal vents and cold seeps on the ocean floor are not dependent on the energy of sunlight. Instead chemosynthetic archaea and bacteria form the base of the food chain.

Origin and Fossil Record

Animals are generally considered to have evolved from a flagellated eukaryote. Their closest known living relatives are the choanoflagellates, collared flagellates that have a morphology similar to the choanocytes of certain sponges. Molecular studies place animals in a super group called the opisthokonts, which also include the choanoflagellates, fungi and a few small parasitic protists. The name comes from the posterior location of the flagellum in motile cells, such as most animal spermatozoa, whereas other eukaryotes tend to have anterior flagella.

The first fossils that might represent animals appear towards the end of the Precambrian, around 610 million years ago, and are known as the Ediacaran or Vendian biota. These are difficult to relate to later fossils, however.

Some may represent precursors of modern phyla, but they may be separate groups, and it is possible they are not really animals at all. Aside from them, most known animal phyla make a more or less simultaneous appearance during the Cambrian period, about 542 million years ago.

It is still disputed whether this event, called the Cambrian explosion, represents a rapid divergence between different groups or a change in conditions that made fossilization possible. However some paleontologists and geologists would suggest that animals appeared much earlier than previously thought, possibly even as early as 1 billion years ago.

Trace fossils such as tracks and burrows found in Tonian era indicate the presence of triploblastic worm like metazoans roughly as large (about 5 mm wide) and complex as earthworms.

In addition during the beginning of the Tonian period around 1 billion years ago (roughly the same time that the trace fossils previously discussed in this article date back to) there was a decrease in Stromatolite diversity which may indicate the appearance of grazing animals during this time as Stromatolites also increased in diversity shortly after the end-Ordovician and end-Permian rendered large amounts of grazing marine animals extinct and decreased shortly after their populations recovered. The discovery that tracks very similar to these early trace fossils are produced today by the giant single-called protist *Gromia sphaerica* casts further doubt on their interpretation as evidence of early animal evolution.

Groups of Animals

Porifera, Radiata and Basal Bilateria

The sponges (Porifera) were long thought to have diverged from other animals early. They lack the complex organization found in most other phyla. Their cells are differentiated, but in most cases not organized into distinct tissues. Sponges typically feed by drawing in water through pores. Archaeocyatha, which have fused skeletons, may represent sponges or a separate phylum. However, a phylogenomic study in 2008 of 150 genes in 21 genera revealed that it is the Ctenophora or comb jellies which are the basal lineage of animals, at least among those 21 phyla. The authors speculate that sponges—or at least those lines of sponges they investigated—are not so primitive, but may instead be secondarily simplified.

Among the other phyla, the Ctenophora and the Cnidaria, which includes sea anemones, corals, and jellyfish, are radially symmetric and have digestive chambers with a single opening, which serves as both the mouth and the anus. Both have distinct tissues, but they are not organized into organs. There are only two main germ layers, the ectoderm and endoderm, with only scattered cells between them. As such, these animals are sometimes called diploblastic. The tiny placozoans are similar, but they do not have a permanent digestive chamber.

The remaining animals form a monophyletic group called the Bilateria. For the most part, they are bilaterally symmetric, and often have a specialized head with feeding and sensory organs. The body is triploblastic, i.e. all three germ layers are well-developed, and tissues form distinct organs. The digestive chamber has two openings, a mouth and an anus, and there is also an internal body cavity called a coelom or pseudocoelom. There are exceptions to each of these characteristics, however — for instance adult echinoderms are radially symmetric, and certain parasitic worms have extremely simplified body structures.

Genetic studies have considerably changed our understanding of the relationships within the Bilateria. Most appear to belong to two major lineages: the deuterostomes and the protostomes, the latter of which includes the Ecdysozoa, Platyzoa, and Lophotrochozoa. In addition, there are a few small groups of bilaterians with relatively similar structure that appear to have diverged before these major groups. These include the Acoelomorpha, Rhombozoa, and Orthonectida. The Myxozoa, single-called parasites that were originally considered Protozoa, are now believed to have developed from the Medusozoa as well.

Deuterostomes

Deuterostomes differ from the other Bilateria, called protostomes, in several ways. In both cases there is a complete digestive tract. However, in protostomes the initial opening (the archenteron) develops into the mouth, and an anus forms separately. In deuterostomes this is reversed. In most protostomes, cells simply fill in the interior of the gastrula to form the mesoderm, called schizocoelous development, but in deuterostomes it forms through invagination of the endoderm, called enterocoelic pouching. Deuterostomes also have a dorsal, rather than a ventral, nerve chord and their embryos undergo different cleavage.

All this suggests the deuterostomes and protostomes are separate, monophyletic lineages. The main phyla of deuterostomes are the Echinodermata and Chordata. The former are radially symmetric and exclusively marine, such as starfish, sea urchins, and sea cucumbers. The latter are dominated by the vertebrates, animals with backbones. These include fish, amphibians, reptiles, birds, and mammals.

In addition to these, the deuterostomes also include the Hemichordata or acorn worms. Although they are not especially prominent today, the important fossil graptolites may belong to this group.

The Chaetognatha or arrow worms may also be deuterostomes, but more recent studies suggest protostome affinities.

Ecdysozoa

The Ecdysozoa are protostomes, named after the common trait of growth by moulting or ecdysis. The largest animal phylum belongs here, the Arthropoda, including insects, spiders, crabs, and their kin. All these organisms have a body divided into repeating segments, typically with paired appendages. Two smaller phyla, the Onychophora and Tardigrada, are close relatives of the arthropods and share these traits.

The ecdysozoans also include the Nematoda or roundworms, perhaps the second largest animal phylum. Roundworms are typically microscopic, and occur in nearly every environment where there is water. A number are important parasites. Smaller phyla related to them are the Nematomorpha or horsehair worms, and the Kinorhyncha, Priapulida, and Loricifera. These groups have a reduced coelom, called a pseudocoelom.

The remaining two groups of protostomes are sometimes grouped together as the Spiralia, since in both embryos develop with spiral cleavage.

Platyzoa

The Platyzoa include the phylum Platyhelminthes, the flatworms. These were originally considered some of the most primitive Bilateria, but it now appears they developed from more complex ancestors. A number of parasites are included in this group, such as the flukes and tapeworms. Flatworms are acoelomates, lacking a body cavity, as are their closest relatives, the microscopic Gastrotricha.

The other platyzoan phyla are mostly microscopic and pseudocoelomate. The most prominent are the Rotifera or rotifers, which are common in aqueous environments. They also include the Acanthocephala or spiny-headed worms, the Gnathostomulida, Micrognathozoa, and possibly the Cycliophora. These groups share the presence of complex jaws, from which they are called the Gnathifera.

Lophotrochozoa

The Lophotrochozoa include two of the most successful animal phyla, the Mollusca and Annelida. The former, which is the second-largest animal phylum by number of described species, includes animals such as snails, clams, and squids, and the latter comprises the segmented worms, such as earthworms and leeches. These two groups have long been considered close relatives because of the common presence of trochophore larvae, but the annelids were considered closer to the arthropods, because they are both segmented. Now this is generally considered convergent evolution, owing to many morphological and genetic differences between the two phyla.

The Lophotrochozoa also include the Nemertea or ribbon worms, the Sipuncula, and several phyla that have a fan of cilia around the mouth, called a lophophore. These were traditionally grouped together as the lophophorates but it now appears they are paraphyletic, some closer to the Nemertea and some to the Mollusca and Annelida. They include the Brachiopoda or lamp shells, which are prominent in the fossil record, the Entoprocta, the Phoronida, and possibly the Bryozoa or moss animals.

Model Organisms

Because of the great diversity found in animals, it is more economical for scientists to study a small number of chosen species so that connections can be drawn from their work and conclusions extrapolated about how animals function in general. Because they are easy to keep and breed, the fruit fly *Drosophila melanogaster* and the nematode *Caenorhabditis elegans* have long been the most intensively studied metazoan model organisms, and were among the first life-forms to be genetically sequenced. This was facilitated by the severely reduced state of their genomes, but the double-edged sword here is that with many genes, introns and linkages lost, these ecdysozoans can teach us little about the origins of animals in general. The extent of this type of evolution within the superphylum will be revealed by the crustacean, annelid, and molluscan genome projects currently in progress. Analysis of the starlet sea anemone genome has emphasised the importance of sponges, placozoans, and choanoflagellates, also being sequenced, in explaining the arrival of 1500 ancestral genes unique to the Eumetazoa. An analysis of the homoscleromorph sponge *Oscarella carmela* also suggests that the last common ancestor of sponges and the eumetazoan animals was more complex than previously assumed.

Other model organisms belonging to the animal kingdom include the mouse (*Mus musculus*) and zebrafish (*Danio rerio*).

Geographical Distribution of Animals: Migration

Animal migration is the travelling of long distances in search of a new habitat. The trigger for the migration may be local climate, local availability of food, or the season of the year. To be counted as a true migration, and not just a local dispersal or irruption, the movement of the animals should be an annual or seasonal occurrence, such as birds migrating south for the winter, or a major habitat change as part of their life cycle, such as young Atlantic salmon leaving the river of their birth when they have reached a few inches in size.

Multiple Generation Migration

In some insect species, such as the monarch butterfly, the whole migration is not carried out by one individual. Instead the butterflies mate and reproduce on the journey, and successive generations travel the next stage of the migration.

Other Types of Animal Migration

Many aquatic animals make a vertical migration, travelling up and down the water column. This is usually a movement of up to a few hundred metres depth and occurs on a daily basis (Diel vertical migration), rather than a seasonal one.

Human Cultural Responses to Animal Migration

Before the phenomenon of animal migration was understood, various folklore and erroneous explanations sprang up to account for the disappearance or sudden arrival of birds in an area. In Ancient Greece, Aristotle proposed that robins turned into redstarts when summer arrived. The barnacle goose was explained in European Medieval bestiaries and manuscripts as either growing like fruit on trees, or developing from goose barnacles on pieces of driftwood. Another example is the swallow, which at various times was suggested to hibernate either underwater, buried in muddy riverbanks, or in hollow trees.

Hibernation and Aestivation

Hibernation

Hibernation is a state of inactivity and metabolic depression in animals, characterized by lower body temperature, slower breathing, and lower metabolic rate. Hibernating animals conserve food, especially during winter when food is short, tapping energy reserves, body fat, at a slow rate. It is the animal's slowed metabolic rate which leads to a reduction in body temperature and not the other way around.

Hibernation may last several days or weeks depending on species, ambient temperature, and time of year, and fur on the animal's body. The typical winter season for a hibernator is characterized by periods of hibernation interrupted by sporadic euthermic arousals wherein body temperature is restored to typical levels. There is a hypothesis that hibernators build a need for sleep during hibernation more slowly than normally, and must occasionally warm up in order to eat. This has been supported by some evidence in the arctic ground squirrel.

Animals that hibernate include bats, some species of ground squirrels and other rodents, mouse lemurs, the European Hedgehog and other insectivores, monotremes and marsupials. Even some rattlesnakes, such as the Western Diamondback, are known to hibernate in caves every winter. Historically, Pliny the Elder believed swallows hibernated, and ornithologist Gilbert White pointed to anecdotal evidence in *The Natural History of Selborne* that indicated as much. Birds typically do not hibernate, instead utilizing torpor. However the Common Poorwill does hibernate. Many experts believe that the processes of daily torpor and hibernation form a continuum.

One animal that some famously consider a hibernator is the bear, although bears do not go into "true hibernation". During a bear's winter sleep state, the degree of metabolic depression is much less than that observed in smaller mammals. Many prefer to use the term "denning". The bear's body temperature remains relatively stable (depressed from 37 °C (99 °F) to approximately 31 °C (88 °F)) and it can be easily aroused. Some reptile species are said to brumate, or undergo brumation, but the connection to this phenomenon with hibernation is not clear.

Hibernating ground squirrels may have abdominal temperatures as low as 0 °C, maintaining sub-zero abdominal temperatures for more than three weeks at a time, although the temperatures at the head and neck remain at 0 C or above. Before entering hibernation most species eat a large amount of food and store energy in fat deposits in order to survive the winter. Some species of mammals hibernate while gestating young, which are born shortly after the mother stops hibernating.

Hibernating animals get their energy by a process known as gluconeogenesis.

For a couple of generations during the 20th century it was thought that basking sharks settled to the floor of the North Sea and hibernated; however, research by Dr. David Sims in 2003 dispelled this hypothesis, showing that the sharks actively travelled huge distances throughout the seasons, tracking the areas with the highest quantity of plankton.

The epaulette sharks have been documented to be able to survive for long periods of time without oxygen, even being left high and dry, and at temperatures of up to 26 °C (79 °F). Other animals able to survive long periods without oxygen include the goldfish, the red-eared slider turtle, the wood frog, and the bar-headed goose.

Until recently no primate, and no tropical mammal, was known to hibernate. However, animal physiologist Kathrin Dausmann of Philipps University of Marburg, Germany, and coworkers presented evidence in the 24 June 2004 edition of *Nature* that the Fat-tailed Dwarf Lemur of Madagascar hibernates in tree holes for seven months of the year. This is interesting because Malagasy winter temperatures sometimes rise to over 30 °C (86 °F), so hibernation is not exclusively an adaptation to low ambient temperatures. The hibernation of this lemur is strongly dependent on the thermal behaviour of its tree hole: if the hole is poorly insulated, the lemur's body temperature fluctuates widely, passively following the ambient temperature; if well insulated, the body temperature stays fairly constant and the animal undergoes regular spells of arousal. Dausmann found that hypometabolism in hibernating animals is not necessarily coupled to a low body temperature.

Noise and vibration from snowmobiles, all-terrain vehicles (ATV) and the like is said to sometimes awaken hibernating animals, who may suffer severely or die as a result of premature awakening in times of food shortage. However, many hibernators can return to hibernation after awakening, and deep hibernators in fact awaken many times throughout the hibernation season in what are called interbout arousals.

Estivation

Estivation or aestivation (from Latin *aestas*, summer), also known as "summer sleep", is a state of animal dormancy somewhat similar to hibernation. It takes place during times of heat and dryness, the hot dry season, which is often but not inevitably the summer months. Invertebrate and vertebrate animals are known to enter this state to avoid damage from high temperatures and the risk of desiccation. Both terrestrial and aquatic animals undergo estivation.

Dormancy

Dormancy is a period in an organism's life cycle when growth, development, and (in animals) physical activity is temporarily stopped. This minimizes metabolic activity and therefore helps an organism to conserve energy. Dormancy tends to be closely associated with environmental conditions. Organisms can synchronize entry to a dormant phase with their environment through predictive or consequential means. Predictive dormancy occurs when an organism enters a dormant phase *before* the onset of adverse conditions. For example, photoperiod and decreasing temperature are used by many plants to predict the onset of winter. Consequential dormancy occurs when organisms enter a dormant phase *after* adverse conditions have arisen. This is commonly found in areas with an unpredictable climate. While very sudden changes in conditions may lead to a high mortality rate among animals relying on consequential dormancy, its use can be advantageous, as organisms remain active longer, and are therefore able to make greater use of available resources.

The Influence of One Organism on the Distribution of Another

Natural Selection

Natural selection is the process by which certain heritable traits—those that make it more likely for an organism to survive and successfully reproduce —become more common in a population over successive generations. It is a key mechanism of evolution.

The natural genetic variation within a population of organisms means that some individuals will survive and reproduce more successfully than others in their current environment. For example, the peppered moth exists in both light and dark colours in the United Kingdom, but during the industrial revolution many of the trees on which the moths rested became blackened by soot, giving the dark-coloured moths an advantage in hiding from predators. This gave dark-coloured moths a better chance of surviving to produce dark-coloured offspring, and in just a few generations the majority of the moths were dark. Factors which affect reproductive success are also important, an issue which Charles Darwin developed in his ideas on sexual selection.

Natural selection acts on the phenotype, or the observable characteristics of an organism, but the genetic (heritable) basis of any phenotype which gives a reproductive advantage will become more common in a population. Over time, this process can result in adaptations that specialize populations for particular ecological niches and may eventually result in the emergence of new species. In other words, natural selection is an important process (though not the only process) by which evolution takes place within a population of organisms. As opposed to artificial selection, in which humans favour specific traits, in natural selection the environment acts as a sieve through which only certain variations can pass.

Natural selection is one of the cornerstones of modern biology. The term was introduced by Darwin in his influential 1859 book *On the Origin of Species*, in which natural selection was described as analogous to artificial selection, a process by which animals and plants with traits considered desirable by human breeders are systematically favoured for reproduction. The concept of natural selection was originally developed in the absence of a valid theory of heredity; at the time of Darwin's writing, nothing was known of modern genetics. The union of traditional Darwinian evolution with subsequent discoveries in classical and molecular genetics is termed the *modern evolutionary synthesis*. Natural selection remains the primary explanation for adaptive evolution.

General Principles

Darwin's illustrations of beak variation in the finches of the Galapagos Islands, which hold 13 closely related species that differ most markedly in the shape of their beaks. The beak of each species is suited to its preferred food, suggesting that beak shapes evolved by natural selection.

Natural variation occurs among the individuals of any population of organisms. Many of these differences do not affect survival (such as differences in eye colour in humans), but some differences may improve the chances of survival of a particular individual. A rabbit that runs faster than others may be more likely to escape from predators, and algae that are more efficient at extracting energy from sunlight will grow faster. Individuals that have better odds for survival also have better odds for reproduction.

If the traits that give these individuals a reproductive advantage are also heritable, that is, passed from parent to child, then there will be a slightly higher proportion of fast rabbits or efficient algae in the next generation. This is known as *differential reproduction*. Even if the reproductive advantage is very slight, over many generations any heritable advantage will become dominant in the population, due to exponential growth. In this way the natural environment of an organism "selects" for traits that confer a reproductive advantage, causing gradual changes or evolution of life. This effect was first described and named by Charles Darwin.

The concept of natural selection predates the understanding of genetics, which is the study of heredity. In modern times, it is understood that selection acts on an organism's phenotype, or observable characteristics, but it is the organism's genetic make-up or genotype that is inherited. The phenotype is the result of the genotype and the environment in which the organism lives. This is the link between natural selection and genetics, as described in the modern evolutionary synthesis. Although a complete theory of evolution also requires an account of how genetic variation arises in the first place (such as by mutation and sexual reproduction) and includes other evolutionary mechanisms (such as gene flow), natural selection is still understood as a fundamental mechanism for evolution.

Nomenclature and Usage

The term *natural selection* has slightly different definitions in different contexts. It is most often defined to operate on heritable traits, because these are the traits that directly participate in evolution. However, natural selection is "blind" in the sense that changes in phenotype (physical and behavioural characteristics) can give a reproductive advantage regardless of whether or not the trait is heritable (non heritable traits can be the result of environmental factors or the life experience of the organism).

Following Darwin's primary usage the term is often used to refer to both the evolutionary consequence of blind selection and to its mechanisms. It is sometimes helpful to explicitly distinguish between selection's mechanisms and its effects; when this distinction is important, scientists define "natural selection" specifically as "those mechanisms that contribute to the selection of individuals that reproduce", without regard to whether the basis of the selection is heritable. This is sometimes referred to as "phenotypic natural selection".

Traits that cause greater reproductive success of an organism are said to be selected for, whereas those that reduce success are selected against. Selection for a trait may also result in the selection of other correlated traits that do not themselves directly influence reproductive advantage. This may occur as a result of pleiotropy or gene linkage.

Fitness

The concept of fitness is central to natural selection. Broadly, individuals which are more "fit" have better potential for survival, as in the well-known phrase "survival of the fittest". However, as with natural selection above, the precise meaning of the term is much more subtle, and Richard Dawkins manages in his later books to avoid it entirely. Modern evolutionary theory defines fitness not by how long an organism lives, but by how successful it is at reproducing. If an organism lives half as long as others of its species, but has twice as many offspring surviving to adulthood, its genes will become more common in the adult population of the next generation.

Though natural selection acts on individuals, the effects of chance mean that fitness can only really be defined "on average" for the individuals within a population. The fitness of a particular genotype corresponds to the average effect on all individuals with that genotype. Very low-fitness genotypes cause their bearers to have few or no offspring on average; examples include many human genetic disorders like cystic fibrosis.

Since fitness is an averaged quantity, it is also possible that a favorable mutation arises in an individual that does not survive to adulthood for unrelated reasons. Fitness also depends crucially upon the environment. Conditions like sickle-cell anemia may have low fitness in the general human population, but because the sickle-cell trait confers immunity from malaria, it has high fitness value in populations which have high malaria infection rates.

Types of Selection

Natural selection can act on any phenotypic trait, and selective pressure can be produced by any aspect of the environment, including sexual selection and competition with members of the same species. However, this does not imply that natural selection is always directional and results in adaptive evolution; natural selection often results in the maintenance of the status quo by eliminating less fit variants.

The unit of selection can be the individual or it can be another level within the hierarchy of biological organisation, such as genes, cells, and kin groups. There is still debate about whether natural selection acts at the level of groups or species to produce adaptations that benefit a larger, non-kin group. Selection at a different level such as the gene can result in an increase in fitness for that gene, while at the same time reducing the fitness of the individuals carrying that gene, in a process called intragenomic conflict. Overall, the combined effect of all selection pressures at various levels determines the overall fitness of an individual, and hence the outcome of natural selection.

Natural selection occurs at every life stage of an individual. An individual organism must survive until adulthood before it can reproduce, and selection of those that reach this stage is called *viability selection.* In many species, adults must compete with each other for mates via sexual selection, and success in this competition determines who will parent the next generation. When individuals can reproduce more than once, a longer survival in the reproductive phase increases the number of offspring, called *survival selection.*

The fecundity of both females and males (for example, giant sperm in certain species of *Drosophila*) can be limited via "fecundity selection". The viability of produced gametes can differ, while intragenomic conflicts such as meiotic drive between the haploid gametes can result in gametic or "genic selection". Finally, the union of some combinations of eggs and sperm might be more compatible than others; this is termed *compatibility selection.*

Sexual Selection

It is useful to distinguish between "ecological selection" and "sexual selection". Ecological selection covers any mechanism of selection as a result of the environment (including relatives, e.g. kin selection, competition, and infanticide), while "sexual selection" refers specifically to competition for mates.

Sexual selection can be *intrasexual,* as in cases of competition among individuals of the same sex in a population, or *intersexual,* as in cases where one sex controls reproductive access by choosing among a population of available mates. Most commonly, intrasexual selection involves male–male competition and intersexual selection involves female choice of suitable males, due to the generally greater investment of resources for a female than a male in a single offspring.

However, some species exhibit sex-role reversed behaviour in which it is males that are most selective in mate choice; the best-known examples of this pattern occur in some fishes of the family *Syngnathidae,* though likely examples have also been found in amphibian and bird species.

Some features that are confined to one sex only of a particular species can be explained by selection exercised by the other sex in the choice of a mate, for example, the extravagant plumage of some male birds. Similarly, aggression between members of the same sex is sometimes associated with very distinctive features, such as the antlers of stags, which are used in combat with other stags. More generally, intrasexual selection is often associated with sexual dimorphism, including differences in body size between males and females of a species.

Examples of Natural Selection

A well-known example of natural selection in action is the development of antibiotic resistance in microorganisms. Since the discovery of penicillin in 1928 by Alexander Fleming, antibiotics have been used to fight bacterial diseases. Natural populations of bacteria contain, among their vast numbers of individual members, considerable variation in their genetic material, primarily as the result of mutations. When exposed to antibiotics, most bacteria die quickly, but some may have mutations that make them slightly less susceptible. If the exposure to antibiotics is short, these individuals will survive the treatment. This selective elimination of maladapted individuals from a population is natural selection.

These surviving bacteria will then reproduce again, producing the next generation. Due to the elimination of the maladapted individuals in the past generation, this population contains more bacteria that have some resistance against the antibiotic. At the same time, new mutations occur, contributing new genetic variation to the existing genetic variation. Spontaneous mutations are very rare, and advantageous mutations are even rarer. However, populations of bacteria are large enough that a few individuals will have beneficial mutations. If a new mutation reduces their susceptibility to an antibiotic, these individuals are more likely to survive when next confronted with that antibiotic.

Given enough time, and repeated exposure to the antibiotic, a population of antibiotic-resistant bacteria will emerge. This new changed population of antibiotic-resistant bacteria is optimally adapted to the context it evolved in. At the same time, it is not necessarily optimally adapted any more to the old antibiotic free environment. The end result of natural selection is two populations that are both optimally adapted to their specific environment, while both perform substandard in the other environment.

The widespread use and misuse of antibiotics has resulted in increased microbial resistance to antibiotics in clinical use, to the point that the methicillin-resistant *Staphylococcus aureus* (MRSA) has been described as a "superbug" because of the threat it poses to health and its relative invulnerability to existing drugs. Response strategies typically include the use of different, stronger antibiotics; however, new strains of MRSA have recently emerged that are resistant even to these drugs.

This is an example of what is known as an evolutionary arms race, in which bacteria continue to develop strains that are less susceptible to antibiotics, while medical researchers continue to develop new antibiotics that can kill them. A similar situation occurs with pesticide resistance in plants and insects. Arms races are not necessarily induced by man; a well-documented example involves the elaboration of the RNA interference pathway in plants as means of innate immunity against viruses.

Evolution by Means of Natural Selection

A prerequisite for natural selection to result in adaptive evolution, novel traits and speciation, is the presence of heritable genetic variation that results in fitness differences. Genetic variation is the result of mutations, recombinations and alterations in the karyotype (the number, shape, size and internal arrangement of the chromosomes). Any of these changes might have an effect that is highly advantageous or highly disadvantageous, but large effects are very rare. In the past, most changes in the genetic material were considered neutral or close to neutral because they occurred in noncoding DNA or resulted in a synonymous substitution. However, recent research suggests that many mutations in non-coding DNA do have slight deleterious effects. Although both mutation rates and average fitness effects of mutations are dependent on the organism, estimates from data in humans have found that a majority of mutations are slightly deleterious.

By the definition of fitness, individuals with greater fitness are more likely to contribute offspring to the next generation, while individuals with lesser fitness are more likely to die early or fail to reproduce. As a result, alleles which on average result in greater fitness become more abundant in the next generation, while alleles which generally reduce fitness become rarer. If the selection forces remain the same for many generations, beneficial alleles become more and more abundant, until they dominate the population, while alleles with a lesser fitness disappear. In every generation, new mutations and recombinations arise spontaneously, producing a new spectrum of phenotypes. Therefore, each new generation will be enriched by the increasing abundance of alleles that contribute to those traits that were favoured by selection, enhancing these traits over successive generations.

Some mutations occur in so-called regulatory genes. Changes in these can have large effects on the phenotype of the individual because they regulate the function of many other genes. Most, but not all, mutations in regulatory genes result in non-viable zygotes. Examples of nonlethal regulatory mutations occur in HOX genes in humans, which can result in a cervical rib or polydactyly, an increase in the number of fingers or toes. When such mutations result in a higher fitness, natural selection will favour these phenotypes and the novel trait will spread in the population.

Established traits are not immutable; traits that have high fitness in one environmental context may be much less fit if environmental conditions change. In the absence of natural selection to preserve such a trait, it will become more variable and deteriorate over time, possibly resulting in a vestigial manifestation of the trait, also called evolutionary baggage. In many circumstances, the apparently vestigial structure may retain a limited functionality, or may be co-opted for other advantageous traits in a phenomenon known as preadaptation. A famous example of a vestigial structure, the eye of the blind mole rat, is believed to retain function in photoperiod perception.

Speciation

Speciation requires selective mating, which result in a reduced gene flow. Selective mating can be the result of 1. Geographic isolation, 2. Behavioural isolation, or 3. Temporal isolation. For example, a change in the physical environment (geographic isolation by an extrinsic barrier) would follow number 1, a change in camouflage for number 2 or a shift in mating times (i.e., one species of deer shifts location and therefore changes its "rut") for number 3. Over time, these subgroups might diverge radically to become different species, either because of differences in selection pressures on the different subgroups, or because different mutations arise spontaneously in the different populations, or because of founder effects – some potentially beneficial alleles may, by chance, be present in only one or other of two subgroups when they first become separated. A lesser-known mechanism of speciation occurs via hybridization, well-documented in plants and occasionally observed in species-rich groups of animals such as cichlid fishes. Such mechanisms of rapid speciation can reflect a mechanism of evolutionary change known as punctuated equilibrium, which suggests that evolutionary change and particularly speciation typically happens quickly after interrupting long periods of stasis.

Genetic changes within groups result in increasing incompatibility between the genomes of the two subgroups, thus reducing gene flow between the groups. Gene flow will effectively cease when the distinctive mutations characterizing each subgroup become fixed. As few as two mutations can result in speciation: if each mutation has a neutral or positive effect on fitness when they occur separately, but a negative effect when they occur together, then fixation of these genes in the respective subgroups will lead to two reproductively isolated populations. According to the biological species concept, these will be two different species.

Historical Development

Pre-Darwinian Theories

Several ancient philosophers expressed the idea that nature produces a huge variety of creatures, apparently randomly, and that only those creatures survive that manage to provide for themselves and reproduce successfully; well-known examples include Empedocles and his intellectual successor, Lucretius, while related ideas were later refined by Aristotle. The struggle for existence was later described by Al-Jahiz, who argued that environmental factors influence animals to develop new characteristics to ensure survival. Abu Rayhan Biruni described the idea of artificial selection and argued that nature works in much the same way. Similar ideas were later expressed by Nasir al-Din Tusi and Ibn Khaldun. Such classical arguments were reintroduced in the 18th century by Pierre Louis Maupertuis and others, including Charles Darwin's grandfather Erasmus Darwin. While these forerunners had an influence on Darwinism, they later had little influence on the trajectory of evolutionary thought after Charles Darwin.

Until the early 19th century, the prevailing view in Western societies was that differences between individuals of a species were uninteresting departures from their Platonic idealism (or typus) of created kinds. However, the theory of uniformitarianism in geology promoted the idea that simple, weak forces could act continuously over long periods of time to produce radical changes in the Earth's landscape. The success of this theory raised awareness of the vast scale of geological time and made plausible the idea that tiny, virtually imperceptible changes in successive generations could produce consequences on the scale of differences between species. Early 19th century evolutionists such as Jean Baptiste Lamarck suggested the inheritance of acquired characteristics as a mechanism for evolutionary change; adaptive traits acquired by an organism during its lifetime could be inherited by that organism's progeny, eventually causing transmutation of species. This theory has come to be known as Lamarckism and was an influence on the anti-genetic ideas of the Stalinist Soviet biologist Trofim Lysenko.

Darwin's Theory

In 1859, Charles Darwin set out his theory of evolution by natural selection as an explanation for adaptation and speciation. He defined natural selection as the "principle by which each slight variation [of a trait], if useful, is preserved". The concept was simple but powerful: individuals best adapted to their environments are more likely to survive and reproduce. As long as there is some variation between them, there will be an inevitable selection of individuals with the most advantageous variations. If the variations are inherited, then differential reproductive success will lead to a progressive evolution of particular populations of a species, and populations that evolve to be sufficiently different eventually become different species.

Darwin's ideas were inspired by the observations that he had made on the *Beagle* voyage, and by the work of two political economists. The first was the Reverend Thomas Malthus, who in *An Essay on the Principle of Population*, noted that population (if unchecked) increases exponentially whereas the food supply grows only arithmetically; thus inevitable limitations of resources would have demographic implications, leading to a "struggle for existence" as a divinely ordained law "in order to rouse man into action, and form his mind to reason" for the greater good despite the "partial evil" limiting population. The second was Adam Smith who, in *The Wealth of Nations*, identified a regulating mechanism in free markets, which he referred to as the "invisible hand", which suggests that prices self-adjust according to supplies and demand. When Darwin read Malthus in 1838 he was already primed by his work as a naturalist to appreciate the "struggle for existence" in nature and it struck him that as population outgrew resources, "favourable variations would tend to be preserved, and unfavourable ones to be destroyed. The result of this would be the formation of new species."

Here is Darwin's own summary of the idea, which can be found in the fourth chapter of the *Origin*:

If during the long course of ages and under varying conditions of life, organic beings vary at all in the several parts of their organisation, and I think this cannot be disputed; if there be, owing to the high geometrical powers of increase of each species, at some age, season, or year, a severe struggle for life, and this certainly cannot be disputed; then, considering the infinite complexity of the relations of all organic beings to each other and to their conditions of existence, causing an infinite diversity in structure, constitution, and habits, to be advantageous to them, I think it would be a most extraordinary fact if no variation ever had occurred useful to each being's own welfare, in the same way as so many variations have occurred useful to man. But if variations useful to any organic being do occur, assuredly individuals thus characterised will have the best chance of being preserved in the struggle for life; and from the strong principle of inheritance they will tend to produce offspring similarly characterised. This principle of preservation, I have called, for the sake of brevity, Natural Selection.

Once he had his theory "by which to work", Darwin was meticulous about gathering and refining evidence as his "prime hobby" before making his idea public. He was in the process of writing his "big book" to present his researches when the naturalist Alfred Russel Wallace independently conceived of the principle and described it in an essay he sent to Darwin to forward to Charles Lyell. Lyell and Joseph Dalton Hooker decided (without Wallace's knowledge) to present his essay together with unpublished writings which Darwin had sent to fellow naturalists, and *On the Tendency of Species to form Varieties; and on the Perpetuation of Varieties and Species by Natural Means of Selection* was read to the Linnean Society announcing co-discovery of the principle in July 1858. Darwin published a detailed account of his evidence and conclusions in *On the Origin of Species* in 1859. In the 3rd edition of 1861 Darwin acknowledged that others — notably William Charles Wells in 1813, and Patrick Matthew in 1831 — had proposed similar ideas, but had neither developed them nor presented them in notable scientific publications.

Darwin thought of natural selection by analogy to how farmers select crops or livestock for breeding, which he called "artificial selection"; in his early manuscripts he referred to a 'Nature' which would do the selection. At the time, other mechanisms of evolution such as evolution by genetic drift were not yet explicitly formulated, and Darwin believed that selection was likely only part of the story: "I am convinced that [it] has been the main, but not exclusive means of modification." In a letter to Charles Lyell in September 1860, Darwin regretted the use of the term "Natural Selection", preferring the term "Natural Preservation". For Darwin and his contemporaries, natural selection was essentially synonymous with evolution by natural selection.

After the publication of *The Origin of Species*, educated people generally accepted that evolution had occurred in some form. However, natural selection remained controversial as a mechanism, partly because it was perceived to be too weak to explain the range of observed characteristics of living organisms, and partly because even supporters of evolution balked at its "unguided" and non-progressive nature, a response that has been characterized as the single most significant impediment to the idea's acceptance. However, some thinkers enthusiastically embraced natural selection; after reading Darwin, Herbert Spencer introduced the term *survival of the fittest*, which became a popular summary of the theory. The fifth edition of *On the Origin of Species* published in 1869 included Spencer's phrase as an alternative to natural selection, with credit given: "But the expression often used by Mr. Herbert Spencer, of the Survival of the Fittest, is more accurate, and is sometimes equally convenient." Although the phrase is still often used by non-biologists, modern biologists avoid it because it is tautological if "fittest" is read to mean "functionally superior" and is applied to individuals rather than considered as an averaged quantity over populations.

Modern Evolutionary Synthesis

Natural selection relies crucially on the idea of heredity, but it was developed long before the basic concepts of genetics. Although the Austrian monk Gregor Mendel, the father of modern genetics, was a contemporary of Darwin's, his work would lie in obscurity until the early 20th century. Only after the integration of Darwin's theory of evolution with a complex statistical appreciation of Gregor Mendel's 're-discovered' laws of inheritance did natural selection become generally accepted by scientists. The work of Ronald Fisher (who developed the required mathematical language and The Genetical Theory of Natural Selection), J.B.S. Haldane (who introduced the concept of the "cost" of natural selection), Sewall Wright (who elucidated the nature of selection and adaptation), Theodosius Dobzhansky (who established the idea that mutation, by creating genetic diversity, supplied the raw material for natural selection: see Genetics and the Origin of Species), William Hamilton (who conceived of kin selection), Ernst Mayr (who recognised the key importance of reproductive isolation for speciation: see Systematics and the Origin of Species) and many others formed the modern evolutionary synthesis. This synthesis cemented natural selection as the foundation of evolutionary theory, where it remains today.

Impact of the Idea

Darwin's ideas, along with those of Adam Smith and Karl Marx, had a profound influence on 19th century thought. Perhaps the most radical claim of the theory of evolution through natural selection is that "elaborately constructed forms, so different from each other, and dependent on each other in so complex a manner" evolved from the simplest forms of life by a few simple principles. This claim inspired some of Darwin's most ardent supporters—and provoked the most profound opposition. The radicalism of natural selection, according to Stephen Jay Gould, lay in its power to "dethrone some of the deepest and most traditional comforts of Western thought". In particular, it challenged long-standing beliefs in such concepts as a special and exalted place for humans in the natural world and a benevolent creator whose intentions were reflected in nature's order and design.

Cell and Molecular Biology

In the 19th century, Wilhelm Roux, a founder of modern embryology, wrote a book entitled « Der Kampf der Teile im Organismus » (The struggle of parts in the organism) in which he suggested that the development of an organism results from a Darwinian competition between the parts of the embryo, occurring at all levels, from molecules to organs. In recent years, a modern version of this theory has been proposed by Jean-Jacques Kupiec. According to this cellular Darwinism, stochasticity at the molecular level generates diversity in cell types whereas cell interactions impose a characteristic order on the developing embryo.

Social and Psychological Theory

The social implications of the theory of evolution by natural selection also became the source of continuing controversy. Friedrich Engels, a German political philosopher and co-originator of the ideology of communism, wrote in 1872 that "Darwin did not know what a bitter satire he wrote on mankind when he showed that free competition, the struggle for existence, which the economists celebrate as the highest historical achievement, is the normal state of the animal kingdom". Interpretation of natural selection as necessarily 'progressive', leading to increasing 'advances' in intelligence and civilisation, was used as a justification for colonialism and policies of eugenics, as well as broader sociopolitical positions now described as Social Darwinism. Konrad Lorenz won the Nobel Prize in Physiology or Medicine in 1973 for his analysis of animal behaviour in terms of the role of natural selection (particularly group selection). However, in Germany in 1940, in writings that he subsequently disowned, he used the theory as a justification for policies of the Nazi state. He wrote "... selection for toughness, heroism, and social utility...must be accomplished by some human institution, if mankind, in default of selective factors, is not to be ruined by domestication-induced degeneracy. The racial idea as the basis of our state has already accomplished much in this respect." Others have developed ideas that human societies and culture evolve by mechanisms that are analogous to those that apply to evolution of species.

More recently, work among anthropologists and psychologists has led to the development of sociobiology and later evolutionary psychology, a field that attempts to explain features of human psychology in terms of adaptation to the ancestral environment.

The most prominent such example, notably advanced in the early work of Noam Chomsky and later by Steven Pinker, is the hypothesis that the human brain is adapted to acquire the grammatical rules of natural language. Other aspects of human behaviour and social structures, from specific cultural norms such as incest avoidance to broader patterns such as gender roles, have been hypothesized to have similar origins as adaptations to the early environment in which modern humans evolved. By analogy to the action of natural selection on genes, the concept of memes – "units of cultural transmission", or culture's equivalents of genes undergoing selection and recombination – has arisen, first described in this form by Richard Dawkins and subsequently expanded upon by philosophers such as Daniel Dennett as explanations for complex cultural activities, including human consciousness. Extensions of the theory of natural selection to such a wide range of cultural phenomena have been distinctly controversial and are not widely accepted.

Information and Systems Theory

In 1922, Alfred Lotka proposed that natural selection might be understood as a physical principle which could be described in terms of the use of energy by a system, a concept that was later developed by Howard Odum as the maximum power principle whereby evolutionary systems with selective advantage maximise the rate of useful energy transformation. Such concepts are sometimes relevant in the study of applied thermodynamics.

The principles of natural selection have inspired a variety of computational techniques, such as "soft" artificial life, that simulate selective processes and can be highly efficient in 'adapting' entities to an environment defined by a specified fitness function. For example, a class of heuristic optimization algorithms known as genetic algorithms, pioneered by John Holland in the 1970s and expanded upon by David E. Goldberg, identify optimal solutions by simulated reproduction and mutation of a population of solutions defined by an initial probability distribution. Such algorithms are particularly useful when applied to problems whose solution landscape is very rough or has many local minima.

Genetic Basis of Natural Selection

The idea of natural selection predates the understanding of genetics. We now have a much better idea of the biology underlying heritability, which is the basis of natural selection.

Genotype and Phenotype

Natural selection acts on an organism's phenotype, or physical characteristics. Phenotype is determined by an organism's genetic make-up and the environment in which the organism lives. Often, natural selection acts on specific traits of an individual, and the terms phenotype and genotype are used narrowly to indicate these specific traits.

When different organisms in a population possess different versions of a gene for a certain trait, each of these versions is known as an allele. It is this genetic variation that underlies phenotypic traits. A typical example is that certain combinations of genes for eye colour in humans which, for instance, give rise to the phenotype of blue eyes. (On the other hand, when all the organisms in a population share the same allele for a particular trait, and this state is stable over time, the allele is said to be *fixed* in that population.)

Some traits are governed by only a single gene, but most traits are influenced by the interactions of many genes. A variation in one of the many genes that contributes to a trait may have only a small effect on the phenotype; together, these genes can produce a continuum of possible phenotypic values.

Directionality of Selection

When some component of a trait is heritable, selection will alter the frequencies of the different alleles, or variants of the gene that produces the variants of the trait. Selection can be divided into three classes, on the basis of its effect on allele frequencies.

Directional selection occurs when a certain allele has a greater fitness than others, resulting in an increase of its frequency. This process can continue until the allele is fixed and the entire population shares the fitter phenotype. It is directional selection that is illustrated in the antibiotic resistance example above.

Far more common is stabilizing selection (which is commonly confused with *purifying selection*), which lowers the frequency of alleles that have a deleterious effect on the phenotype – that is, produce organisms of lower fitness. This process can continue until the allele is eliminated from the population. Purifying selection results in functional genetic features, such as protein-coding genes or regulatory sequences, being conserved over time due to selective pressure against deleterious variants.

Finally, a number of forms of balancing selection exist, which do not result in fixation, but maintain an allele at intermediate frequencies in a population. This can occur in diploid species (that is, those that have two pairs of chromosomes) when heterozygote individuals, who have different alleles on each chromosome at a single genetic locus, have a higher fitness than homozygote individuals that have two of the same alleles. This is called heterozygote advantage or over dominance, of which the best-known example is the malarial resistance observed in heterozygous humans who carry only one copy of the gene for sickle cell anemia. Maintenance of allelic variation can also occur through disruptive or diversifying selection, which favours genotypes that depart from the average in either direction (that is, the opposite of over dominance), and can result in a bimodal distribution of trait values. Finally, balancing selection can occur through frequency-dependent selection, where the fitness of one particular phenotype depends on the distribution of other phenotypes in the population. The principles of game theory have been applied to understand the fitness distributions in these situations, particularly in the study of kin selection and the evolution of reciprocal altruism.

Selection and Genetic Variation

A portion of all genetic variation is functionally neutral in that it produces no phenotypic effect or significant difference in fitness;

the hypothesis that this variation accounts for a large fraction of observed genetic diversity is known as the neutral theory of molecular evolution and was originated by Motoo Kimura. When genetic variation does not result in differences in fitness, selection cannot *directly* affect the frequency of such variation. As a result, the genetic variation at those sites will be higher than at sites where variation does influence fitness. However, after a period with no new mutation, the genetic variation at these sites will be eliminated due to genetic drift.

Mutation Selection Balance

Natural selection results in the reduction of genetic variation through the elimination of maladapted individuals and consequently of the mutations that caused the maladaptation. At the same time, new mutations occur, resulting in a mutation-selection balance. The exact outcome of the two processes depends both on the rate at which new mutations occur and on the strength of the natural selection, which is a function of how unfavourable the mutation proves to be. Consequently, changes in the mutation rate or the selection pressure will result in a different mutation-selection balance.

Genetic Linkage

Genetic linkage occurs when the loci of two alleles are *linked*, or in close proximity to each other on the chromosome. During the formation of gametes, recombination of the genetic material results in reshuffling of the alleles. However, the chance that such a reshuffle occurs between two alleles depends on the distance between those alleles; the closer the alleles are to each other, the less likely it is that such a reshuffle will occur. Consequently, when selection targets one allele, this automatically results in selection of the other allele as well; through this mechanism, selection can have a strong influence on patterns of variation in the genome.

Selective sweeps occur when an allele becomes more common in a population as a result of positive selection. As the prevalence of one allele increases, linked alleles can also become more common, whether they are neutral or even slightly deleterious. This is called *genetic hitchhiking*. A strong selective sweep results in a region of the genome where the positively selected haplotype (the allele and its neighbours) are essentially the only ones that exist in the population.

Whether a selective sweep has occurred or not can be investigated by measuring linkage disequilibrium, or whether a given haplotype is over represented in the population. Normally, genetic recombination results in a reshuffling of the different alleles within a haplotype, and none of the haplotypes will dominate the population. However, during a selective sweep, selection for a specific allele will also result in selection of neighbouring alleles. Therefore, the presence of strong linkage disequilibrium might indicate that there has been a 'recent' selective sweep, and this can be used to identify sites recently under selection.

Background selection is the opposite of a selective sweep. If a specific site experiences strong and persistent purifying selection, linked variation will tend to be weeded out along with it, producing a region in the genome of low overall variability. Because background selection is a result of deleterious new mutations, which can occur randomly in any haplotype, it produces no linkage disequilibrium.

8

Mountain

A mountain is a large landform that stretches above the surrounding land in a limited area usually in the form of a peak. A mountain is generally steeper than a hill. Mountains are formed through tectonic forces or volcanism. These forces can locally raise the surface of the earth by over 10,000 feet (3,000 m). Mountains erode slowly through the action of rivers, weather conditions, and glaciers. A few mountains are isolated summits, but most occur in huge mountain ranges. The highest mountain on Earth is Mount Everest in the Himalayas of Asia, whose summit is 8,848 m (29,029 ft) above mean sea level. The highest known mountain in the Solar System is Olympus Mons on the planet Mars at 21,171 m (69,459 ft).

Definition

There is no universally accepted definition of a mountain. Elevation, volume, relief, steepness, spacing and continuity have been used as criteria for defining a mountain. In the Oxford English Dictionary a mountain is defined as "a natural elevation of the earth surface rising more or less abruptly from the surrounding level and attaining an altitude which, relatively to the adjacent elevation, is impressive or notable."

Whether a landform is called a mountain may depend on usage among the local people. The highest point in San Francisco, California, is called Mount Davidson, notwithstanding its height of 300 m (980 ft), which makes it ten feet short of the minimum for a mountain by American designations. Similarly, Mount Scott outside Lawton, Oklahoma is only 251 m (823 ft) from its base to its highest point. Whittow's *Dictionary of Physical Geography* states *"Some authorities regard eminences above 600 m (2,000 ft) as mountains, those below being referred to as hills.”*

Definitions of “mountain” include:

- Height over base of at least 2,500 m (8,202 ft);
- Height over base of 1,500 m (4,921 ft).–2,500 m (8,202 ft). with a slope greater than 2 degrees
- Height over base of 1,000 m (3,281 ft).–1,500 m (4,921 ft). with a slope greater than 5 degrees
- Local (radius 7,000 m (22,966 ft). elevation greater than 300 m (984 ft)., or 300 m (984 ft)–1,000 m (3,281 ft). if local (radius 7,000 m (22,966 ft). elevation is greater than 300 m (984 ft).

By these definitions, mountains cover 64% of Asia, 25% of Europe, 22% of South America, 17% of Australia, and 3% of Africa. As a whole, 24% of the Earth’s land mass is mountainous and 10% of people live in mountainous regions. Most of the world’s rivers are fed from mountain sources, and more than half of humanity depends on mountains for water.

Characteristics

Tall mountains reach into the colder layers of the atmosphere. They are consequently subject to glaciation, and erosion through frost action. Such processes produce the peak shape. Some mountains have glacial lakes, created by melting glaciers; for example, there are an estimated 3,000 glacial lakes in Bhutan. Mountains can be eroded and weathered, altering their characteristics over time.

Tall mountains have different climatic conditions at the top than at the base, and will thus have altitudinal zonation of ecosystems. At the highest elevations, trees cannot grow, and whatever life may be present will be of the alpine type, resembling tundra. Just below the tree line, one may find subalpine forests of needleleaf trees, which can withstand cold, dry conditions. In regions with dry climates, the tendency of mountains to have higher precipitation as well as lower temperatures also provides for varying conditions, which in turn leads to differing flora and fauna. Some plants and animals found in these zones tend to become isolated since the conditions above and below a particular zone will be inhospitable and thus constrain their movements or dispersal. On the other hand, birds, being capable of flight, may take advantage of montane habitats and migrate into a region that would otherwise not provide appropriate habitat. These isolated ecological systems, or microclimates, are known as sky islands.

Mountains are generally colder than their surrounding lowlands due to the way that the sun heats the surface of the Earth. Practically all the heat at the surface of the Earth comes from the sun, in the form of solar energy. The sun's radiation is absorbed by land and sea, whence the heat is transferred into the air. Static air is a poor conductor of heat, so conduction of heat from the ground to the atmosphere is negligible. Heat is mainly transferred into the atmosphere through convection and radiation. The air immediately adjacent to the warmed surface will rise due to its buoyancy, leading to convective circulation, in the form of thermals, within the lowest layer of the atmosphere, the troposphere.

When heat radiates from the surface of the earth, it is released as long-wave radiation, which can move freely through gases composed of diatomic molecules (such as the atmosphere's oxygen and nitrogen), but is readily absorbed by triatomic molecules, such as carbon dioxide and water vapour. Since most of the atmosphere's quantity of such triatomic gases is contained within the troposphere, this portion of the atmosphere is readily heated by the earth's radiation. The tropopause forms a blanket of air keeping the surface warm. This is the Greenhouse Effect. The higher the altitude, the less of this blanket there is to keep in the heat.

Thus, higher elevations, such as mountains, are colder than surrounding lowlands. Air temperature in the troposphere decreases with gains in altitude. The rate at which the temperature drops with elevation, called the environmental lapse rate, is not constant (it can fluctuate throughout the day or seasonally and also regionally), but a normal lapse rate is 5.5°C per 1,000 m (3.57°F per 1,000 ft). The temperature continues to drop with increasing altitude, until the tropopause (11,000m or 36,089 ft in the U.S. Standard Atmosphere, where it does not decrease further. However, this is higher than the highest mountaintop.

Mountains are generally less preferable for human habitation than lowlands; the weather is often harsher, and there is little level ground suitable for agriculture. The decreasing atmospheric pressure means that less oxygen is available for breathing, and there is less protection against solar radiation (UV). Acute mountain sickness (caused by hypoxia—a lack of oxygen in the blood) affects over half of lowlanders who spend more than a few hours above 3,500 metres (11,480 ft). Many mountains and mountain ranges throughout the world have been left in their natural state, and are today primarily used for recreation, while others are used for logging, mining, grazing.

Some mountains offer spectacular views from their summits, while others are densely wooded.

Summit accessibility is affected by height, steepness, latitude, terrain, weather. Roads, ski lifts, or aerial tramways allow access. Hiking, backpacking, mountaineering, rock climbing, ice climbing, downhill skiing, and snowboarding are recreational activities enjoyed on mountains. Mountains that support heavy recreational use (especially downhill skiing) are often the locations of mountain resorts. Mountains are made up of earth and rock materials. The outermost layer of the Earth or the Earth's crust is composed of seven primary plates. When two plates move or collide with each other, vast land areas are uplifted, forming mountains.

Geology

A mountain is usually produced by the movement of lithospheric plates, either orogenic movement or epeirogenic movement. Compressional forces, isostatic uplift and intrusion of igneous matter forces surface rock upward, creating a landform higher than the surrounding features. The height of the feature makes it either a hill or, if higher and steeper, a mountain. The absolute heights of features termed mountains and hills vary greatly according to an area's terrain. The major mountains tend to occur in long linear arcs, indicating tectonic plate boundaries and activity. Two types of mountain are formed in this way depending on how the rock reacts to the tectonic forces, — fold mountains or fault-block mountains. Other mountain building processes include volcanoes and sea floor spreading.

Fold Mountains

Compressional forces in continental collisions may cause the compressed region to thicken and fold, with material forced both upwards and downwards. Since the less dense continental crust "floats" (cf iceberg) on the denser mantle rocks beneath, the weight of any crustal material forced *upward* to form hills, plateaus or mountains must be balanced by the buoyancy force of a much greater volume forced *downward* into the mantle.

Thus the continental crust is normally much thicker under mountains (sometimes called "mountain roots"), compared to lower lying areas. However, in many continental collisions (e.g. the Himalayas) part of one continent may simply *override* the other, crumpling in the process with the overridden crust forming much of the support.

Mountains may similarly be partly supported by oceanic crust subducted beneath the continental crust (e.g. the Andes as the Nazca plate flows beneath the South American Plate).

Fault-block Mountains

Block mountains are created when large areas are widely broken up by faults creating large vertical displacements. This occurrence is fairly common. The uplifted blocks are block mountains or *horsts*. The intervening dropped blocks are termed *graben*: these can be small or form extensive rift valley systems. This form of landscape can be seen in East Africa, the Vosges, the Basin and Range province of Western North America and the Rhine valley. These areas often occur when the regional stress is extensional and the crust is thinned.

Rock that does not fault may fold, either symmetrically or asymmetrically. The upfolds are *anticlines* and the downfolds are *synclines*: in asymmetric folding there may also be recumbent and overturned folds. The Jura Mountains are an example of folding. Over time, erosion can bring about an inversion of relief: the soft upthrust rock is worn away so the anticlines are actually lower than the tougher, more compressed rock of the synclines.

Volcanoes

Some isolated mountains are produced by volcanoes, including many apparently small islands or seamounts that reach a great height above the ocean floor.

Mid-ocean Ridges

The mid-ocean ridges formed during sea-floor spreading are often referred to as undersea mountain ranges due to their bathymetric prominence.

Plain

In geography, a plain is land with relatively low relief, that is flat or gently rolling. Prairies and steppes are types of plains, and the archetype for a plain is often thought of as a grassland, but plains in their natural state may also be covered in shrublands, woodland and forest, or vegetation may be absent in the case of sandy or stony plains in hot deserts. Types of flatlands for which the term is not generally used include those covered entirely and permanently by swamps, marshes, playas, or ice sheets.

Plains occur as lowlands and at the bottoms of valleys but also on plateaus at high elevations. In a valley, a plain is enclosed on two sides but in other cases a plain may be delineated by a complete or partial ring of hills, by mountains or cliffs. Where a geological region contains more than one plain, they may be connected by a pass (sometime termed a gap). Plains may have been formed from flowing lava, deposited by water, ice or wind, or formed by erosion by these agents from hills and mountains.

Plains in many areas are important for agriculture because where the soils were deposited as sediments they may be deep and fertile, and the flatness facilitates mechanization of crop production; or because they support grasslands which provide good grazing for livestock.

Types of Terrestrial Plains

- Coastal plain, an area of low-lying land adjacent to a sea; the term is used especially where they contrast with hills, mountains or plateau further inland.
- Alluvial plains are formed by rivers, and may be one of these overlapping types:
    - o Flood plain, adjacent to a stream, river, lake or wetland that experiences occasional or periodic flooding.
    - o Alluvial plain, formed over a long period of time by a river depositing sediment on its floodplain or bed which becomes alluvial soil. The difference between a floodplain and an alluvial plain is that the floodplain represents the area experiencing flooding fairly regularly in the present or recently, whereas an alluvial plain includes areas where the floodplain is now and used to be, or areas which only experience flooding a few times a century.
    - o Scroll plain, a plain through which a river meanders with a very low gradient.
- Lacustrine plain, a plain that originally formed in a lacustrine environment, that is, as the bed of a lake.
- Lava plain, formed by sheets of flowing lava.
- Glacial plains are formed by the movement of glaciers under the force of gravity:
    - o Till plain, a plain of glacial till that forms when a sheet of ice becomes detached from the main body of a glacier and melts in place depositing the sediments it carries. Till plains are composed of unsorted material (till) of all sizes.
    - o Sandur (plural sandar), a glacial out-wash plain formed of sediments deposited by melt-water at the terminus of a

glacier. Sandar consist mainly of stratified (layered and sorted) gravel and sand

- o Mud plain Mud plains are usually found close to swamps and quicksand. Mud plains are usually found near the jungle or lakes.
- o Grassland plain Grassland plains are formed by moss and algae, which look like grasslands.
- o Old Believings It was believed that in England in the 14th century, that plains of any kind is where the dead wandered searching for their revenge. This was thought tho be true because so many people went missing when wandering the plains at night and were never heard of again.

Lake

A lake is a body of relatively still water of considerable size, localized in a basin, that is surrounded by land apart from a river, stream, or other form of moving water that serves to feed or drain the lake. Lakes are inland and not part of the ocean and therefore are distinct from lagoons, and are larger and deeper than ponds. Lakes can be contrasted with rivers or streams, which are usually flowing. However most lakes are fed and drained by rivers and streams.

Natural lakes are generally found in mountainous areas, rift zones, and areas with ongoing glaciation. Other lakes are found in endorheic basins or along the courses of mature rivers. In some parts of the world there are many lakes because of chaotic drainage patterns left over from the last Ice Age. All lakes are temporary over geologic time scales, as they will slowly fill in with sediments or spill out of the basin containing them. Many lakes are artificial and are constructed for industrial or agricultural use, for hydro-electric power generation or domestic water supply, or for aesthetic or recreational purposes.

Etymology, Meaning and Usage of "Lake"

The word *lake* comes from Middle English *lake* ("lake, pond, waterway"), from Old English *lacu* ("pond, pool, stream"), from Proto-Germanic *lakô* ("pond, ditch, slow moving stream"), from the Proto-Indo-European root *leõ-* ("to leak, drain"). Cognates include Dutch *laak* ("lake, pond, ditch"), Middle Low German *lâke* ("water pooled in a riverbed, puddle"), German *Lache* ("pool, puddle"), and Icelandic *lækur* ("slow flowing stream"). Also related are the English words *leak* and *leach*. There is considerable uncertainty about defining the difference between lakes and ponds, and no current internationally accepted definition of either term across scientific disciplines or political boundaries exists. For example, limnologists have defined lakes as water bodies which are simply a larger version of a pond, which can have wave action on the shoreline or where wind-induced turbulence plays a major role in mixing the water column.

None of these definitions completely excludes ponds and all are difficult to measure. For this reason there has been increasing use made of simple size-based definitions to separate ponds and lakes. One definition of *lake* is a body of water of 2 hectares (5 acres) or more in area, however others have defined lakes as waterbodies of 5 hectares (12 acres) and above, or 8 hectares (20 acres) and above.

Charles Elton, one of the founders of ecology, regarded lakes as waterbodies of 40 hectares (99 acres) or more. The term *lake* is also used to describe a feature such as Lake Eyre, which is a dry basin most of the time but may become filled under seasonal conditions of heavy rainfall.

In common usage many lakes bear names ending with the word *pond*, and a lesser number of names ending with *lake* are in quasi-technical fact, ponds. One textbook illustrates this point with the following: "In Newfoundland, for example, almost every lake is called a pond, whereas in Wisconsin, almost every pond is called a lake."

One hydrology book proposes to define it as a body of water with the following five chacteristics:

- it partially or totally fills one or several basins connected by straits
- has essentially the same water level in all parts (except for relatively short-lived variations caused by wind, varying ice cover, large inflows, etc.)
- it does not have regular intrusion of sea water
- a considerable portion of the sediment suspended in the water is captured by the basins (for this to happen they need to have a sufficiently small inflow-to-volume ratio)
- the area measured at the mean water level exceeds an arbitrarily chosen threshold (for instance, one hectare)

With the exception of the sea water intrusion criterion, the other ones have been accepted or elaborated upon by other hydrology publications.

Distribution of Lakes

The majority of lakes on Earth are fresh water, and most lie in the Northern Hemisphere at higher latitudes. Canada, with a deranged drainage system has an estimated 63,504 lakes. Finland has 187,888 lakes of which 60,000 are large. Most lakes have at least one natural outflow in the form of a river or stream, which maintain a lake's average level by allowing the drainage of excess water. Some lakes do not have a natural outflow and lose water solely by evapouration or underground seepage or both. They are termed endorheic lakes. Many lakes are artificial and are constructed for hydro-electric power generation, aesthetic purposes, recreational purposes, industrial use, agricultural use or domestic water supply. Evidence of extraterrestrial lakes exists; "definitive evidence of lakes filled with methane" was announced by NASA as returned by the Cassini Probe observing the moon Titan, which orbits the planet Saturn.

Globally, lakes are greatly outnumbered by ponds: of an estimated 304 million standing water bodies worldwide, 91% are 1 hectare (2.5 acres) or less in area. Small lakes are also much more numerous than large lakes: in terms of area, one-third of the world's standing water is represented by lakes and ponds of 10 hectares (25 acres) or less. However, large lakes account for much of the area of standing water with 122 large lakes of 1,000 square kilometres (390 sq mi, 100,000 ha, 247,000 acres) or more representing about 29% of the total global area of standing inland water.

Origin of Natural Lakes

There are a number of natural processes that can form lakes. A recent tectonic uplift of a mountain range can create bowl-shaped depressions that accumulate water and form lakes. The advance and retreat of glaciers can scrape depressions in the surface where water accumulates; such lakes are common in Scandinavia, Patagonia, Siberia and Canada. The most notable examples are probably the Great Lakes of North America. Lakes can also form by means of landslides or by glacial blockages. An example of the latter occurred during the last ice age in the U.S. state of Washington, when a huge lake formed behind a glacial flow; when the ice retreated, the result was an immense flood that created the Dry Falls at Sun Lakes, Washington.

Salt lakes (also called saline lakes) can form where there is no natural outlet or where the water evapourates rapidly and the drainage surface of the water table has a higher-than-normal salt content. Examples of salt lakes include Great Salt Lake, the Aral Sea and the Dead Sea. Small, crescent-shaped lakes called oxbow lakes can form in river valleys as a result of meandering. The slow-moving river forms a sinuous shape as the outer side of bends are eroded away more rapidly than the inner side. Eventually a horseshoe bend is formed and the river cuts through the narrow neck. This new passage then forms the main passage for the river and the ends of the bend become silted up, thus forming a bow-shaped lake.

Crater lakes are formed in volcanic craters and calderas which fill up with precipitation more rapidly than they empty via evapouration. Sometimes the latter are called caldera lakes, although often no distinction is made. An example is Crater Lake in Oregon, located within the caldera of Mount Mazama. The caldera was created in a massive volcanic eruption that led to the subsidence of Mount Mazama around 4860 BC. Gloe Lakes are freshwater lakes that have emerged when the water they consists of has been separated, not considerably long before, from the sea as a consequence of post-glacial rebound. Some lakes, such as Lake Jackson in Florida, USA, come into existence as a result of sinkhole activity. Lake Vostok is a subglacial lake in Antarctica, possibly the largest in the world. The pressure from the ice atop it and its internal chemical composition mean that, if the lake were drilled into, a fissure could result that would spray somewhat like a geyser.

Most lakes are geologically young and shrinking since the natural results of erosion will tend to wear away the sides and fill the basin. Exceptions are those such as Lake Baikal and Lake Tanganyika that lie along continental rift zones and are created by the crust's subsidence as two plates are pulled apart. These lakes are the oldest and deepest in the world. Lake Baikal, which is 25-30 million years old, is deepening at a faster rate than it is being filled by erosion and may be destined over millions of years to become attached to the global ocean. The Red Sea, for example, is thought to have originated as a rift valley lake.

Characteristics

Lakes have numerous features in addition to lake type, such as drainage basin (also known as catchment area), inflow and outflow, nutrient content, dissolved oxygen, pollutants, pH, and sedimentation. Changes in the level of a lake are controlled by the difference between the input and output compared to the total volume of the lake. Significant input sources are precipitation onto the lake, runoff carried by streams and channels from the lake's catchment area, groundwater channels and aquifers, and artificial sources from outside the catchment area. Output sources are evapouration from the lake, surface and groundwater flows, and any extraction of lake water by humans. As climate conditions and human water requirements vary, these will create fluctuations in the lake level.

Lakes can be also categorized on the basis of their richness in nutrients, which typically affect plant growth. Nutrient-poor lakes are said to be *oligotrophic* and are generally clear, having a low concentration of plant life. *Mesotrophic lakes* have good clarity and an average level of nutrients. *Eutrophic* lakes are enriched with nutrients, resulting in good plant growth and possible algal blooms. *Hypertrophic* lakes are bodies of water that have been excessively enriched with nutrients.

These lakes typically have poor clarity and are subject to devastating algal blooms. Lakes typically reach this condition due to human activities, such as heavy use of fertilizers in the lake catchment area. Such lakes are of little use to humans and have a poor ecosystem due to decreased dissolved oxygen.

Due to the unusual relationship between water's temperature and its density, lakes form layers called thermoclines, layers of drastically varying temperature relative to depth. Fresh water is most dense at about 4 degrees Celsius (39.2 °F) at sea level.

When the temperature of the water at the surface of a lake reaches the same temperature as deeper water, as it does during the cooler months in temperate climates, the water in the lake can mix, bringing oxygen-starved water up from the depths and bringing oxygen down to decomposing sediments. Deep temperate lakes can maintain a reservoir of cold water year-round, which allows some cities to tap that reservoir for deep lake water cooling.

Deeper layer becomes oxygen starved and can become saturated with carbon dioxide, or other gases such as sulfur dioxide if there is even a trace of volcanic activity. Exceptional events, such as earthquakes or landslides, can cause mixing which rapidly brings the deep layers up to the surface and release a vast cloud of gas which lay trapped in solution in the colder water at the bottom of the lake. This is called a limnic eruption.

An example is the disaster at Lake Nyos in Cameroon. The amount of gas that can be dissolved in water is directly related to pressure. As deep water surfaces, the pressure drops and a vast amount of gas comes out of solution. Under these circumstances carbon dioxide is hazardous because it is heavier than air and displaces it, so it may flow down a river valley to human settlements and cause mass asphyxiation. The material at the bottom of a lake, or *lake bed*, may be composed of a wide variety of inorganics, such as silt or sand, and organic material, such as decaying plant or animal matter. The composition of the lake bed has a significant impact on the flora and fauna found within the lake's environs by contributing to the amounts and the types of nutrients available. A paired (black and white) layer of the varved lake sediments correspond to a year. During winter, when organisms die, carbon is deposited down, resulting to a black layer. At the same year, during summer, only few organic materials are deposited, resulting to a white layer at the lake bed. These are commonly used to track past paleontological events.

Limnology

Limnology is the study of inland bodies of water and related ecosystems. Limnology divides lakes into three zones: the *littoral zone*, a sloped area close to land; the *photic* or *open-water zone*, where sunlight is abundant; and the deep-water *profundal* or *benthic zone*, where little sunlight can reach. The depth to which light can reach in lakes depends on turbidity, determined by the density and size of suspended particles. A particle is in suspension if its weight is less than the random turbidity forces acting upon it. These particles can be sedimentary or biological in origin and are responsible for the colour of the water. Decaying plant matter, for instance, may be responsible for a yellow or brown colour, while algae may cause greenish water. In very shallow water bodies, iron oxides make water reddish brown. Biological particles include algae and detritus. Bottom-dwelling detritivorous fish can be responsible for turbid waters, because they stir the mud in search of food. Piscivorous fish contribute to turbidity by eating plant-eating (planktonivorous) fish, thus increasing the amount of algae.

The light depth or transparency is measured by using a *Secchi disk*, a 20-cm (8 in) disk with alternating white and black quadrants. The depth at which the disk is no longer visible is the *Secchi depth*, a measure of transparency. The Secchi disk is commonly used to test for eutrophication. A lake moderates the surrounding region's temperature and climate because water has a very high specific heat capacity (4,186 $J \cdot kg^{-1} \cdot K^{-1}$). In the daytime a lake can cool the land beside it with local winds, resulting in a sea breeze; in the night it can warm it with a land breeze.

How Lakes Disappear

The lake may be infilled with deposited sediment and gradually become a wetland such as a swamp or marsh. Large water plants, typically totos, accelerate this closing process significantly because they partially decompose to form peat soils that fill the shallows. Conversely, peat soils in a marsh can naturally burn and reverse this process to recreate a shallow lake. Turbid lakes and lakes with many plant-eating fish tend to disappear more slowly.

A "disappearing" lake (barely noticeable on a human timescale) typically has extensive plant mats at the water's edge. These become a new habitat for other plants, like peat moss when conditions are right, and animals, many of which are very rare. Gradually the lake closes and young peat may form, forming a fen. In lowland river valleys where a river can meander, the presence of peat is explained by the infilling of historical oxbow lakes. In the very last stages of succession, trees can grow in, eventually turning the wetland into a forest.

Some lakes can disappear seasonally. These are called intermittent lakes, ephemeral lakes, or seasonal lakes and can be found in karstic terrain. A prime example of an intermittent lake is Lake Cerknica in Slovenia or Lag Prau Pulte in Graubünden. Other intermittent lakes are only the result of above-average precipitation in a closed, or endorheic basin, usually filling dry lake beds. This can occur in some of the driest places on earth, like Death Valley. This occurred in the spring of 2005, after unusually heavy rains. The lake did not last into the summer, and was quickly evapourated. A more commonly filled lake of this type is Sevier Lake of west-central Utah.

Sometimes a lake will disappear quickly. On 3 June 2005, in Nizhny Novgorod Oblast, Russia, a lake called Lake Beloye vanished in a matter of minutes. News sources reported that government officials theorized that this strange phenomenon may have been caused by a shift in the soil underneath the lake that allowed its water to drain through channels leading to the Oka River. The presence of ground permafrost is important to the persistence of some lakes. According to research published in the journal *Science* ("Disappearing Arctic Lakes", June 2005), thawing permafrost may explain the shrinking or disappearance of hundreds of large Arctic lakes across western Siberia. The idea here is that rising air and soil temperatures thaw permafrost, allowing the lakes to drain away into the ground.

Some lakes disappear because of human development factors. The shrinking Aral Sea is described as being "murdered" by the diversion for irrigation of the rivers feeding it.

Extraterrestrial Lakes

At present the surface of the planet Mars is too cold and has too little atmospheric pressure to permit the pooling of liquid water on the surface. Geologic evidence appears to confirm, however, that ancient lakes once formed on the surface. It is also possible that volcanic activity on Mars will occasionally melt subsurface ice, creating large lakes. Under current conditions this water would quickly freeze and sublimate unless insulated in some manner, such as by a coating of volcanic ash.

Only one world other than Earth is known to harbour lakes, Saturn's largest moon, Titan. Photographs and spectroscopic analysis by the Cassini–Huygens spacecraft show liquid ethane on the surface, which is thought to be mixed with liquid methane. Jupiter's small moon Io is volcanically active due to tidal stresses, and as a result sulfur deposits have accumulated on the surface. Some photographs taken during the Galileo mission appear to show lakes of liquid sulfur on the surface. There are dark basaltic plains on the Moon, similar to lunar maria but smaller, that are called *lacus* (singular *lacus*, Latin for "lake") because they were thought by early astronomers to be lakes of water.

Notable Lakes

- The deepest lake is Lake Baikal in Siberia, with a bottom at 1,637 m. Its mean depth is also the greatest in the world (749 m).

  It is also the world's largest lake by volume (23,600 km^3, though smaller than the Caspian Sea at 78,200 km^3), and the second longest (about 630 km from tip to tip).
- The longest lake is Lake Tanganyika, with a length of about 660 km (measured along the lake's center line).

  It is also the second largest by volume and second deepest (1,470 m) in the world, after lake Baikal.
- The world's oldest lake is Lake Baikal, followed by Lake Tanganyika (Tanzania).
- The world's highest lake, if size is not a criterion, may be the crater lake of Ojos del Salado, at 6,390 metres (20,965 ft).
- The highest large (greater than 250 square kilometres) freshwater lake in the world is the 290 km^2 Pumoyong Tso (Pumuoyong Tso), in the Tibet Autonomous Region of China, at 28-34N 90-24E, 5,018 metres (16,463 ft) above sea level.

- The world's highest commercially navigable lake is Lake Titicaca in Peru and Bolivia at 3,812 m (12,507 ft). It is also the largest freshwater (and second largest overall) lake in South America.
- The world's lowest lake is the Dead Sea, bordering Israel and Jordan at 418 m (1,371 ft) below sea level. It is also one of the lakes with highest salt concentration.
- Lake Huron has the longest lake coastline in the world: about 2980 km, excluding the coastline of its many inner islands.
- The largest island in a freshwater lake is Manitoulin Island in Lake Huron, with a surface area of 2,766 km². Lake Manitou, located on Manitoulin Island, is the largest lake on an island in a freshwater lake.
- The largest lake located on an island is Nettilling Lake on Baffin Island, with an area of 5,542 km² and a maximum length of 123 km.
- The largest lake in the world that drains naturally in two directions is Wollaston Lake.
- Lake Toba on the island of Sumatra is located in what is probably the largest resurgent caldera on Earth.
- The largest lake located completely within the boundaries of a single city is Lake Wanapitei in the city of Sudbury, Ontario, Canada. Before the current city boundaries came into effect in 2001, this status was held by Lake Ramsey, also in Sudbury.
- Lake Enriquillo in Dominican Republic is the only saltwater lake in the world inhabited by crocodiles.
- Lake Bernard, Ontario, Canada, claims to be the largest freshwater lake in the world with no islands.
- The largest lake in one country is Lake Michigan, in the U.S.A. However, it is sometimes considered part of Lake Michigan-Huron, making the record go to Great Bear Lake, Northwest Territories, in Canada, the largest lake within one jurisdiction.
- The largest lake on an island in a lake on an island is Crater Lake on Vulcano Island in Lake Taal on the island of Luzon, The Philippines.
- The northernmost named lake on Earth is Upper Dumbell Lake in the Qikiqtaaluk Region of Nunavut, Canada at a latitude of 82°28'N. It is located only 5.2 kilometres southwest of Alert, the northernmost settlement in the world. There are

also several small lakes located north of Upper Dumbell Lake, but they are all unnamed and only appear on very detailed maps.

Fluvial

Fluvial is a term used in geography and Earth science to refer to the processes associated with rivers and streams and the deposits and landforms created by them. When the stream or rivers are associated with glaciers, ice sheets, or ice caps, the term glaciofluvial or fluvioglacial is used.

Fluvial processes comprise the motion of sediment and erosion or deposition (geology) on the river bed. Erosion by moving water can happen in two ways. Firstly, the movement of water across the bed exerts a shear stress directly onto the bed. If the cohesive strength of the substrate is lower than the shear exerted, or the bed is composed of loose sediment which can be mobilized by such stresses, then the bed will be lowered purely by clearwater flow. However, if the river carries significant quantities of sediment, this material can act as tools to enhance wear of the bed (abrasion). At the same time the fragments themselves are ground down, becoming smaller and more rounded (attrition).

Sediment in rivers is transported as either bedload (the coarser fragments which move close to the bed) or suspended load (finer fragments carried in the water). There is also a component carried as dissolved material. For each grain size there is a specific velocity at which the grains start to move, called *entrainment velocity*. However the grains will continue to be transported even if the velocity falls below the entrainment velocity due to the reduced (or removed) friction between the grains and the river bed. Eventually the velocity will fall low enough for the grains to be deposited. This is shown by the Hjulstrøm curve.

A river is continually picking up and dropping solid particles of rock and soil from its bed throughout its length. Where the river flow is fast, more particles are picked up than dropped. Where the river flow is slow, more particles are dropped than picked up. Areas where more particles are dropped are called alluvial or flood plains, and the dropped particles are called alluvium. Even small streams make alluvial deposits, but it is in the flood plains and deltas of large rivers that large, geologically-significant alluvial deposits are found.

The amount of matter carried by a large river is enormous. The names of many rivers derive from the colour that the transported matter gives the water. For example, the Huang He in China is literally translated "Yellow River", and the Mississippi River in the United States is also called "the Big Muddy." It has been estimated that the Mississippi River annually carries 406 million tons of sediment to the sea, the Huang He 796 million tons, and the Po River in Italy 67 million tons.

9

Rock

In geology, a rock is a naturally occurring solid aggregate of one or more minerals or mineraloids. For example, the common rock, granite, is a combination of the quartz, feldspar and biotite minerals. The Earth's outer solid layer, the lithosphere, is made of rock. Rocks have been used by mankind through out history. From the Stone Age rocks have been used for tools. The minerals and metals we find in rocks have been essential to human civilization. Three major groups of rocks are defined: igneous, sedimentary, and metamorphic. The scientific study of rocks is called petrology, which is an essential component of geology.

Rock Classification

Rocks are generally classified by mineral and chemical composition, by the texture of the constituent particles and by the processes that formed them. These indicators separate rocks into three types: igneous, sedimentary, and metamorphic. They are further classified according to particle size. The transformation of one rock type to another is described by the geological model called the rock cycle.

Igneous rocks are formed when molten magma cools and are divided into two main categories: plutonic rock and volcanic. Plutonic or intrusive rocks result when magma cools and crystallizes slowly within the Earth's crust (example granite), while volcanic or extrusive rocks result from magma reaching the surface either as lava or fragmental ejecta (examples pumice and basalt).

Sedimentary rocks are formed by deposition of either clastic sediments, organic matter, or chemical precipitates (evapourites), followed by compaction of the particulate matter and cementation during diagenesis. Sedimentary rocks form at or near the Earth's surface. Mud rocks comprise 65% (mudstone, shale and siltstone); sandstones 20 to 25% and carbonate rocks 10 to 15% (limestone and dolostone). Metamorphic rocks are formed by subjecting any rock type (including previously formed metamorphic rock) to different temperature and pressure conditions than those in which the original rock was formed. These temperatures and pressures are always higher than those at the Earth's surface and must be sufficiently high so as to change the original minerals into other mineral types or else into other forms of the same minerals (e.g. by recrystallization).

The three classes of rocks—the igneous, the sedimentary and the metamorphic—are subdivided into many groups. There are, however, no hard and fast boundaries between allied rocks. By increase or decrease in the proportions of their constituent minerals they pass by every gradation into one another, the distinctive structures also of one kind of rock may often be traced gradually merging into those of another. Hence the definitions adopted in establishing rock nomenclature merely correspond to selected points (more or less arbitrary) in a continuously graduated series.

Igneous Rocks

Igneous rock (derived from the Latin word *igneus* meaning of fire, from *ignis* meaning fire) forms through the cooling and solidification of magma or lava. Igneous rock may form with or without crystallization, either below the surface as intrusive (plutonic) rocks or on the surface as extrusive (volcanic) rocks. This magma can be derived from partial melts of pre-existing rocks in either a planet's mantle or crust. Typically, the melting is caused by one or more of three processes: an increase in temperature, a decrease in pressure, or a change in composition. Over 700 types of igneous rocks have been described, most of them having formed beneath the surface of Earth's crust. These have diverse properties, depending on their composition and how they were formed.

Sedimentary Rocks

Sedimentary rocks are types of rock that are formed by the deposition of material at the Earth's surface and within bodies of water. Sedimentation is the collective name for processes that cause mineral and/or organic particles (detritus) to settle and accumulate or for minerals to precipitate from a solution. Particles that form a sedimentary rock by accumulating are called sediment. Before being deposited, sediment was formed by weathering and erosion in a source area, and then transported to the place of deposition by water, wind, ice, mass movement or glaciers which are called agents of denudation.

Metamorphic Rocks

Metamorphic rocks arise from the transformation of existing rock types, in a process called metamorphism, which means "change in form". The original rock (protolith) is subjected to heat and pressure, (temperatures greater than 150 to 200 °C and pressures of 1500 bars) causing profound physical and/or chemical change. The protolith may be sedimentary rock, igneous rock or another older metamorphic rock.

Human Use

The use of rocks has had a huge impact on the cultural and technological development of the human race. Rocks have been used by humans and other hominids for more than 2 million years. Lithic technology marks some of the oldest and continuously used technologies. The mining of rocks for their metal ore content has been one of the most important factors of human advancement, which has progressed at different rates in different places in part because of the kind of metals available from the rocks of a region.

Mining

Mining is the extraction of valuable minerals or other geological materials from the earth, from an ore body, vein or (coal) seam. This term also includes the removal of soil. Materials recovered by mining include base metals, precious metals, iron, uranium, coal, diamonds, limestone, oil shale, rock salt and potash. Mining is required to obtain any material that cannot be grown through agricultural processes, or created artificially in a laboratory or factory. Mining in a wider sense comprises extraction of any non-renewable resource (e.g., petroleum, natural gas, or even water).

Mining of stone and metal has been done since pre-historic times. Modern mining processes involve prospecting for ore bodies, analysis of the profit potential of a proposed mine, extraction of the desired materials and finally reclamation of the land to prepare it for other uses once the mine is closed. The nature of mining processes creates a potential negative impact on the environment both during the mining operations and for years after the mine is closed.

This impact has led to most of the world's nations adopting regulations to moderate the negative effects of mining operations.

Safety has long been a concern as well, though modern practices have improved safety in mines significantly.

Volcano

A volcano is an opening, or rupture, in a planet's surface or crust, which allows hot magma, volcanic ash and gases to escape from below the surface. Volcanoes are generally found where tectonic plates are diverging or converging. A mid-oceanic ridge, for example the Mid-Atlantic Ridge, has examples of volcanoes caused by divergent tectonic plates pulling apart; the Pacific Ring of Fire has examples of volcanoes caused by convergent tectonic plates coming together.

By contrast, volcanoes are usually not created where two tectonic plates slide past one another. Volcanoes can also form where there is stretching and thinning of the Earth's crust in the interiors of plates, e.g., in the East African Rift, the Wells Gray-Clearwater volcanic field and the Rio Grande Rift in North America.

This type of volcanism falls under the umbrella of "Plate hypothesis" volcanism. Volcanism away from plate boundaries has also been explained as mantle plumes.

These so-called "hotspots", for example Hawaii, are postulated to arise from upwelling diapirs with magma from the core–mantle boundary, 3,000 km deep in the Earth. Erupting volcanoes can pose many hazards, not only in the immediate vicinity of the eruption.

Volcanic ash can be a threat to aircraft, in particular those with jet engines where ash particles can be melted by the high operating temperature. Large eruptions can affect temperature as ash and droplets of sulfuric acid obscure the sun and cool the Earth's lower atmosphere or troposphere; however, they also absorb heat radiated up from the Earth, thereby warming the stratosphere.

Historically, so-called volcanic winters have caused catastrophic famines.

Plate Tectonics

Divergent Plate Boundaries

At the mid-oceanic ridges, two tectonic plates diverge from one another. New oceanic crust is being formed by hot molten rock slowly cooling and solidifying. The crust is very thin at mid-oceanic ridges due to the pull of the tectonic plates. The release of pressure due to the thinning of the crust leads to adiabatic expansion, and the partial melting of the mantle causing volcanism and creating new oceanic crust. Most divergent plate boundaries are at the bottom of the oceans, therefore most volcanic activity is submarine, forming new seafloor. Black smokers or deep sea vents are an example of this kind of volcanic activity. Where the mid-oceanic ridge is above sea-level, volcanic islands are formed, for example, Iceland.

Convergent Plate Boundaries

Subduction zones are places where two plates, usually an oceanic plate and a continental plate, collide. In this case, the oceanic plate subducts, or submerges under the continental plate forming a deep ocean trench just offshore. Water released from the subducting plate lowers the melting temperature of the overlying mantle wedge, creating magma. This magma tends to be very viscous due to its high silica content, so often does not reach the surface and cools at depth. When it does reach the surface, a volcano is formed. Typical examples for this kind of volcano are Mount Etna and the volcanoes in the Pacific Ring of Fire.

"Hotspots"

"Hotspots" is the name given to volcanic provinces postulated to be formed by mantle plumes. These are postulated to comprise columns of hot material that rise from the core-mantle boundary. They are suggested to be hot, causing large-volume melting, and to be fixed in space. Because the tectonic plates move across them, each volcano becomes dormant after a while and a new volcano is then formed as the plate shifts over the postulated plume. The Hawaiian Islands have been suggested to have been formed in such a manner, as well as the Snake River Plain, with the Yellowstone Caldera being the part of the North American plate currently above the hot spot. This theory is currently under criticism, however.

Volcanic Features

The most common perception of a volcano is of a conical mountain, spewing lava and poisonous gases from a crater at its summit. This describes just one of many types of volcano, and the features of volcanoes are much more complicated.

The structure and behaviour of volcanoes depends on a number of factors. Some volcanoes have rugged peaks formed by lava domes rather than a summit crater, whereas others present landscape features such as massive plateaus. Vents that issue volcanic material (lava, which is what magma is called once it has escaped to the surface, and ash) and gases (mainly steam and magmatic gases) can be located anywhere on the landform. Many of these vents give rise to smaller cones.

Other types of volcano include cryovolcanoes (or ice volcanoes), particularly on some moons of Jupiter, Saturn and Neptune; and mud volcanoes, which are formations often not associated with known magmatic activity. Active mud volcanoes tend to involve temperatures much lower than those of igneous volcanoes, except when a mud volcano is actually a vent of an igneous volcano.

Shield Volcanoes

Shield volcanoes, so named for their broad, shield-like profiles, are formed by the eruption of low-viscosity lava that can flow a great distance from a vent. They generally do not explode catastrophically. Since low-viscosity magma is typically low in silica, shield volcanoes are more common in oceanic than continental settings. The Hawaiian volcanic chain is a series of shield cones, and they are common in Iceland, as well.

Lava Domes

Lava domes are built by slow eruptions of highly viscous lavas. They are sometimes formed within the crater of a previous volcanic eruption (as in Mount Saint Helens), but can also form independently, as in the case of Lassen Peak. Like stratovolcanoes, they can produce violent, explosive eruptions, but their lavas generally do not flow far from the originating vent.

Volcanic Cones (Cinder Cones)

Volcanic cones or cinder cones result from eruptions of mostly small pieces of scoria and pyroclastics (both resemble cinders, hence the name of this volcano type) that build up around the vent. These can be relatively short-lived eruptions that produce a cone-shaped hill perhaps 30 to 400 metres high. Most cinder cones erupt only once. Cinder cones may form as flank vents on larger volcanoes, or occur on their own. Parícutin in Mexico and Sunset Crater in Arizona are examples of cinder cones. In New Mexico, Caja del Rio is a volcanic field of over 60 cinder cones.

Stratovolcanoes (Composite Volcanoes)

Stratovolcanoes or composite volcanoes are tall conical mountains composed of lava flows and other ejecta in alternate layers, the strata that give rise to the name. Stratovolcanoes are also known as composite volcanoes, created from several structures during different kinds of eruptions. Strato/composite volcanoes are made of cinders, ash and lava. Cinders and ash pile on top of each other, lava flows on top of the ash, where it cools and hardens, and then the process begins again. Classic examples include Mt. Fuji in Japan, Mayon Volcano in the Philippines, and Mount Vesuvius and Stromboli in Italy.

Throughout recorded history, ash produced by the explosive eruption of stratovolcanoes has posed the greatest hazard to civilizations as compared to other types of volcanoes. No supervocano has erupted in human history. Shield volcanos have smaller pressure buildup from the underlying lava flow as compared to stratovolcanoes. Fissure vents and monogenetic volcanic fields (volcanic cones) have less powerful eruptions, as they are many times under extension. Stratovolcanoes have been a greater historical threat because they are steeper than shield volcanos, with slopes of 30–35° compared to slopes of generally 5–10°, and their loose tephra are material for dangerous lahars.

Supervolcanoes

A supervolcano is a large volcano that usually has a large caldera and can potentially produce devastation on an enormous, sometimes continental, scale. Such eruptions would be able to cause severe cooling of global temperatures for many years afterwards because of the huge volumes of sulfur and ash erupted. They are the most dangerous type of volcano. Examples include Yellowstone Caldera in Yellowstone National Park and Valles Caldera in New Mexico (both western United States), Lake Taupo in New Zealand, Lake Toba in Sumatra, Indonesia and Ngorogoro Crater in Tanzania, Krakatoa near Java and Sumatra, Indonesia. Supervolcanoes are hard to identify centuries later, given the enormous areas they cover. Large igneous provinces are also considered supervolcanoes because of the vast amount of basalt lava erupted, but are non-explosive.

Submarine Volcanoes

Submarine volcanoes are common features on the ocean floor. Some are active and, in shallow water, disclose their presence by blasting steam and rocky debris high above the surface of the sea. Many others lie at such great depths that the tremendous weight of the water above them prevents the explosive release of steam and gases, although they can be detected by hydrophones and discolouration of water because of volcanic gases. Pumice rafts may also appear.

Even large submarine eruptions may not disturb the ocean surface. Because of the rapid cooling effect of water as compared to air, and increased buoyancy, submarine volcanoes often form rather steep pillars over their volcanic vents as compared to above-surface volcanoes. They may become so large that they break the ocean surface as new islands. Pillow lava is a common eruptive product of submarine volcanoes. Hydrothermal vents are common near these volcanoes, and some support peculiar ecosystems based on dissolved minerals.

Subglacial volcanoes develop underneath icecaps. They are made up of flat lava which flows at the top of extensive pillow lavas and palagonite. When the icecap melts, the lavas on the top collapse, leaving a flat-topped mountain. These volcanoes are also called table mountains, tuyas or (uncommonly) mobergs. Very good examples of this type of volcano can be seen in Iceland, however, there are also tuyas in British Columbia. The origin of the term comes from Tuya Butte, which is one of the several tuyas in the area of the Tuya River and Tuya Range in northern British Columbia. Tuya Butte was the first such landform analyzed and so its name has entered the geological literature for this kind of volcanic formation. The Tuya Mountains Provincial Park was recently established to protect this unusual landscape, which lies north of Tuya Lake and south of the Jennings River near the boundary with the Yukon Territory.

Mud Volcanoes

Mud volcanoes or mud domes are formations created by geo-excreted liquids and gases, although there are several processes which may cause such activity. The largest structures are 10 kilometres in diametre and reach 700 metres high.

Erupted Material

Lava Composition

Another way of classifying volcanoes is by the *composition of material erupted* (lava), since this affects the shape of the volcano. Lava can be broadly classified into 4 different compositions (Cas & Wright, 1987):

- If the erupted magma contains a high percentage (>63%) of silica, the lava is called felsic.
  - o Felsic lavas (dacites or rhyolites) tend to be highly viscous (not very fluid) and are erupted as domes or short, stubby flows. Viscous lavas tend to form stratovolcanoes or lava domes. Lassen Peak in California is an example of a

volcano formed from felsic lava and is actually a large lava dome.

- o Because siliceous magmas are so viscous, they tend to trap volatiles (gases) that are present, which cause the magma to erupt catastrophically, eventually forming stratovolcanoes. Pyroclastic flows (ignimbrites) are highly hazardous products of such volcanoes, since they are composed of molten volcanic ash too heavy to go up into the atmosphere, so they hug the volcano's slopes and travel far from their vents during large eruptions. Temperatures as high as 1,200 °C are known to occur in pyroclastic flows, which will incinerate everything flammable in their path and thick layers of hot pyroclastic flow deposits can be laid down, often up to many metres thick. Alaska's Valley of Ten Thousand Smokes, formed by the eruption of Novarupta near Katmai in 1912, is an example of a thick pyroclastic flow or ignimbrite deposit. Volcanic ash that is light enough to be erupted high into the Earth's atmosphere may travel many kilometres before it falls back to ground as a tuff.

- If the erupted magma contains 52–63% silica, the lava is of *intermediate* composition.
    - o These "andesitic" volcanoes generally only occur above subduction zones (e.g. Mount Merapi in Indonesia).
    - o Andesitic lava is typically formed at convergent boundary margins of tectonic plates, by several processes:
        - § Melting of subducted slab containing sediments
        - § Magma mixing between felsic rhyolitic and mafic basaltic magmas in an intermediate reservoir prior to emplacement or lava flow.
- If the erupted magma contains <52% and >45% silica, the lava is called mafic (because it contains higher percentages of magnesium (Mg) and iron (Fe)) or basaltic. These lavas are usually much less viscous than rhyolitic lavas, depending on their eruption temperature; they also tend to be hotter than felsic lavas. Mafic lavas occur in a wide range of settings:
    - o At mid-ocean ridges, where two oceanic plates are pulling apart, basaltic lava erupts as pillows to fill the gap;
    - o Shield volcanoes (e.g. the Hawaiian Islands, including Mauna Loa and Kilauea), on both oceanic and continental crust;

- o As continental flood basalts.
- Some erupted magmas contain <=45% silica and produce ultramafic lava. Ultramafic flows, also known as komatiites, are very rare; indeed, very few have been erupted at the Earth's surface since the Proterozoic, when the planet's heat flow was higher. They are (or were) the hottest lavas, and probably more fluid than common mafic lavas.

Volcanic Activity

Popular Classification of Volcanoes

A popular way of classifying magmatic volcanoes is by their frequency of eruption, with those that erupt regularly called active, those that have erupted in historical times but are now quiet called dormant or inactive, and those that have not erupted in historical times called extinct. However, these popular classifications—extinct in particular—are practically meaningless to scientists. They use classifications which refer to a particular volcano's formative and eruptive processes and resulting shapes, which was explained above.

Active

There is no consensus among volcanologists on how to define an "active" volcano. The lifespan of a volcano can vary from months to several million years, making such a distinction sometimes meaningless when compared to the lifespans of humans or even civilizations. For example, many of Earth's volcanoes have erupted dozens of times in the past few thousand years but are not currently showing signs of eruption. Given the long lifespan of such volcanoes, they are very active. By human lifespans, however, they are not.

Scientists usually consider a volcano to be erupting or likely to erupt if it is currently erupting, or showing signs of unrest such as unusual earthquake activity or significant new gas emissions. Most scientists consider a volcano *active* if it has erupted in the last 10,000 years (Holocene times) – the Smithsonian Global Volcanism Programme uses this definition of *active*. There are about 1500 active volcanoes in the world – the majority along the Pacific Ring of Fire – and around 50 of these erupt each year. An estimated 500 million people live near active volcanoes.

Historical times (that is, in recorded history) is another timeframe for *active*. The *Catalogue of the Active Volcanoes of the World*, published by the International Association of Volcanology, uses this definition, by which there are more than 500 active volcanoes. However the span of recorded history differs from region to region. In China and the Mediterranean, it reaches back nearly 3,000 years, but in the Pacific Northwest of the United States and Canada, it reaches back less than 300 years, and in Hawaii and New Zealand, only around 200 years.

Extinct

Extinct volcanoes are those that scientists consider unlikely to erupt again, because the volcano no longer has a magma supply. Examples of extinct volcanoes are many volcanoes on the Hawaiian – Emperor seamount chain in the Pacific Ocean, Hohentwiel, Shiprock and the Zuidwal volcano in the Netherlands. Edinburgh Castle in Scotland is famously located atop an extinct volcano.

Otherwise, whether a volcano is truly extinct is often difficult to determine. Since "supervolcano" calderas can have eruptive lifespans sometimes measured in millions of years, a caldera that has not produced an eruption in tens of thousands of years is likely to be considered dormant instead of extinct. Some volcanologists refer to extinct volcanoes as inactive, though the term is now more commonly used for dormant volcanoes once thought to be extinct.

Dormant

It is difficult to distinguish an extinct volcano from a dormant (inactive) one. Volcanoes are often considered to be extinct if there are no written records of its activity. Nevertheless, volcanoes may remain dormant for a long period of time. For example, Yellowstone has a repose/recharge period of around 700 ka, and Toba of around 380 ka. Vesuvius was described by Roman writers as having been covered with gardens and vineyards before its famous eruption of AD 79, which destroyed the towns of Herculaneum and Pompeii.

Before its catastrophic eruption of 1991, Pinatubo was an inconspicuous volcano, unknown to most people in the surrounding areas. Two other examples are the long-dormant Soufrière Hills volcano on the island of Montserrat, thought to be extinct before activity resumed in 1995 and Fourpeaked Mountain in Alaska, which, before its September 2006 eruption, had not erupted since before 8000 BC and had long been thought to be extinct.

Technical Classification of Volcanoes

Volcanic-alert Level

The three common popular classifications of volcanoes can be subjective and some volcanoes thought to have been extinct have erupted again. To help prevent people from falsely believing they are not at risk when living on or near a volcano, countries have adopted new classifications to describe the various levels and stages of volcanic activity. Some alert systems use different numbers or colours to designate the different stages. Other systems use colours and words. Some systems use a combination of both.

Volcano Warning Schemes of the United States

The United States Geological Survey (USGS) has adopted a common system nationwide for characterizing the level of unrest and eruptive activity at volcanoes. The new volcano alert-level system classifies volcanoes now as being in a normal, advisory, watch or warning stage. Additionally, colours are used to denote the amount of ash produced. Details of the US system can be found at Volcano warning schemes of the United States.

Effects of Volcanoes

There are many different types of volcanic eruptions and associated activity: phreatic eruptions (steam-generated eruptions), explosive eruption of high-silica lava (e.g., rhyolite), effusive eruption of low-silica lava (e.g., basalt), pyroclastic flows, lahars (debris flow) and carbon dioxide emission. All of these activities can pose a hazard to humans. Earthquakes, hot springs, fumaroles, mud pots and geysers often accompany volcanic activity.

The concentrations of different volcanic gases can vary considerably from one volcano to the next. Water vapour is typically the most abundant volcanic gas, followed by carbon dioxide and sulfur dioxide. Other principal volcanic gases include hydrogen sulfide, hydrogen chloride, and hydrogen fluoride. A large number of minor and trace gases are also found in volcanic emissions, for example hydrogen, carbon monoxide, halocarbons, organic compounds, and volatile metal chlorides.

Large, explosive volcanic eruptions inject water vapour (H_2O), carbon dioxide (CO_2), sulfur dioxide (SO_2), hydrogen chloride (HCl), hydrogen fluoride (HF) and ash (pulverized rock and pumice) into the stratosphere to heights of 16–32 kilometres (10–20 mi) above the Earth's surface. The most significant impacts from these injections come from the conversion of sulfur dioxide to sulfuric acid (H_2SO_4), which condenses rapidly in the stratosphere to form fine sulfate aerosols. The aerosols increase the Earth's albedo—its reflection of radiation from the Sun back into space – and thus cool the Earth's lower atmosphere or troposphere; however, they also absorb heat radiated up from the Earth, thereby warming the stratosphere. Several eruptions during the past century have caused a decline in the average temperature at the Earth's surface of up to half a degree (Fahrenheit scale) for periods of one to three years – sulfur dioxide from the eruption of Huaynaputina probably caused the Russian famine of 1601–1603.

One proposed volcanic winter happened c. 70,000 years ago following the supereruption of Lake Toba on Sumatra island in Indonesia. According to the Toba catastrophe theory to which some anthropologists and archeologists subscribe, it had global consequences, killing most humans then alive and creating a population bottleneck that affected the genetic inheritance of all humans today. The 1815 eruption of Mount Tambora created global climate anomalies that became known as the "Year Without a Summer" because of the effect on North American and European weather. Agricultural crops failed and livestock died in much of the Northern Hemisphere, resulting in one of the worst famines of the 19th century. The freezing winter of 1740–41, which led to widespread famine in northern Europe, may also owe its origins to a volcanic eruption.

It has been suggested that volcanic activity caused or contributed to the End-Ordovician, Permian-Triassic, Late Devonian mass extinctions, and possibly others. The massive eruptive event which formed the Siberian Traps, one of the largest known volcanic events of the last 500 million years of Earth's geological history, continued for a million years and is considered to be the likely cause of the "Great Dying" about 250 million years ago, which is estimated to have killed 90% of species existing at the time.

The sulfate aerosols also promote complex chemical reactions on their surfaces that alter chlorine and nitrogen chemical species in the stratosphere. This effect, together with increased stratospheric chlorine levels from chlorofluorocarbon pollution, generates chlorine monoxide (ClO), which destroys ozone (O_3). As the aerosols grow and coagulate, they settle down into the upper troposphere where they serve as nuclei for cirrus clouds and further modify the Earth's radiation balance. Most of the hydrogen chloride (HCl) and hydrogen fluoride (HF) are dissolved in water droplets in the eruption cloud and quickly fall to the ground as acid rain. The injected ash also falls rapidly from the stratosphere; most of it is removed within several days to a few weeks. Finally, explosive volcanic eruptions release the greenhouse gas carbon dioxide and thus provide a deep source of carbon for biogeochemical cycles.

Gas emissions from volcanoes are a natural contributor to acid rain. Volcanic activity releases about 130 to 230 teragrams (145 million to 255 million short tons) of carbon dioxide each year. Volcanic eruptions may inject aerosols into the Earth's atmosphere. Large injections may cause visual effects such as unusually colourful sunsets and affect global climate mainly by cooling it. Volcanic eruptions also provide the benefit of adding nutrients to soil through the weathering process of volcanic rocks. These fertile soils assist the growth of plants and various crops. Volcanic eruptions can also create new islands, as the magma cools and solidifies upon contact with the water.

Ash thrown into the air by eruptions can present a hazard to aircraft, especially jet aircraft where the particles can be melted by the high operating temperature. Dangerous encounters in 1982 after the eruption of Galunggung in Indonesia, and 1989 after the eruption of Mount Redoubt in Alaska raised awareness of this phenomenon. Nine Volcanic Ash Advisory Centers were established by the International Civil Aviation Organization to monitor ash clouds and advise pilots accordingly. The 2010 eruptions of Eyjafjallajökull caused major disruptions to air travel in Europe.

Volcanoes on Other Planetary Bodies

The Earth's Moon has no large volcanoes and no current volcanic activity, although recent evidence suggests it may still possess a partially molten core. However, the Moon does have many volcanic features such as maria (the darker patches seen on the moon), rilles and domes. The planet Venus has a surface that is 90% basalt, indicating that volcanism played a major role in shaping its surface. The planet may have had a major global resurfacing event about 500 million years ago, from what scientists can tell from the density of impact craters on the surface.

Lava flows are widespread and forms of volcanism not present on Earth occur as well. Changes in the planet's atmosphere and observations of lightning have been attributed to ongoing volcanic eruptions, although there is no confirmation of whether or not Venus is still volcanically active. However, radar sounding by the Magellan probe revealed evidence for comparatively recent volcanic activity at Venus's highest volcano Maat Mons, in the form of ash flows near the summit and on the northern flank.

There are several extinct volcanoes on Mars, four of which are vast shield volcanoes far bigger than any on Earth. They include Arsia Mons, Ascraeus Mons, Hecates Tholus, Olympus Mons, and Pavonis Mons. These volcanoes have been extinct for many millions of years, but the European *Mars Express* spacecraft has found evidence that volcanic activity may have occurred on Mars in the recent past as well. Jupiter's moon Io is the most volcanically active object in the solar system because of tidal interaction with Jupiter. It is covered with volcanoes that erupt sulfur, sulfur dioxide and silicate rock, and as a result, Io is constantly being resurfaced. Its lavas are the hottest known anywhere in the solar system, with temperatures exceeding 1,800 K (1,500 °C). In February 2001, the largest recorded volcanic eruptions in the solar system occurred on Io. Europa, the smallest of Jupiter's Galilean moons, also appears to have an active volcanic system, except that its volcanic activity is entirely in the form of water, which freezes into ice on the frigid surface. This process is known as cryovolcanism, and is apparently most common on the moons of the outer planets of the solar system.

In 1989 the Voyager 2 spacecraft observed cryovolcanoes (ice volcanoes) on Triton, a moon of Neptune, and in 2005 the Cassini–Huygens probe photographed fountains of frozen particles erupting from Enceladus, a moon of Saturn. The ejecta may be composed of water, liquid nitrogen, dust, or methane compounds. Cassini–Huygens also found evidence of a methane-spewing cryovolcano on the Saturnian moon Titan, which is believed to be a significant source of the methane found in its atmosphere.

It is theorized that cryovolcanism may also be present on the Kuiper Belt Object Quaoar. A 2010 study of the exoplanet COROT-7b, which was detected by transit in 2009, studied that tidal heating from the host star very close to the planet and neighbouring planets could generate intense volcanic activity similar to Io.

10

Soil Science

Soil science is the study of soil as a natural resource on the surface of the earth including soil formation, classification and mapping; physical, chemical, biological, and fertility properties of soils; and these properties in relation to the use and management of soils. Sometimes terms which refer to branches of soil science, such as pedology (formation, chemistry, morphology and classification of soil) and edaphology (influence of soil on organisms, especially plants), are used as if synonymous with soil science.

The diversity of names associated with this discipline is related to the various associations concerned. Indeed, engineers, agronomists, chemists, geologists, physical geographers, ecologists, biologists, microbiologists, sylviculturists, sanitarians, archaeologists, and specialists in regional planning, all contribute to further knowledge of soils and the advancement of the soil sciences. Soil scientists have raised concerns about how to preserve soil and arable land in a world with a growing population, possible future water crisis, increasing per capita food consumption, and land degradation.

Fields of Study

Soil occupies the pedosphere, one of Earth's spheres that the geosciences use to organize the Earth conceptually. This is the conceptual perspective of pedology and edaphology, the two main branches of soil science. Pedology is the study of soil in its natural setting. Edaphology is the study of soil in relation to soil-dependent uses. Both branches apply a combination of soil physics, soil chemistry, and soil biology. Due to the numerous interactions between the biosphere, atmosphere and hydrosphere that are hosted within the pedosphere, more integrated, less soil-centric concepts are also valuable. Many concepts essential to understanding soil come from individuals not identifiable strictly as soil scientists. This highlights the interdisciplinary nature of soil concepts.

Research

Dependence on and curiosity about soil, exploring the diversity and dynamic of this resource continues to yield fresh discoveries and insights. New avenues of soil research are compelled by a need to understand soil in the context of climate change, greenhouse gases, and carbon sequestration. Interest in maintaining the planet's biodiversity and in exploring past cultures has also stimulated renewed interest in achieving a more refined understanding of soil.

Mapping

Most empirical knowledge of soil in nature comes from soil survey efforts. Soil survey, or soil mapping, is the process of determining the soil types or other properties of the soil cover over a landscape, and mapping them for others to understand and use. It relies heavily on distinguishing the individual influences of the five classic soil forming factors. This effort draws upon geomorphology, physical geography, and analysis of vegetation and land-use patterns. Primary data for the soil survey are acquired by field sampling and supported by remote sensing.

Classification

As of 2006, the World Reference Base for Soil Resources, via its Land & Water Development division, is the pre-eminent soil classification system. It replaces the previous FAO soil classification. The WRB borrows from modern soil classification concepts, including USDA soil taxonomy. The classification is based mainly on soil morphology as an expression pedogenesis. A major difference with USDA soil taxonomy is that soil climate is not part of the system, except insofar as climate influences soil profile characteristics.

Many other classification schemes exist, including vernacular systems. The structure in vernacular systems are either nominal, giving unique names to soils or landscapes, or descriptive, naming soils by their characteristics such as red, hot, fat, or sandy. Soils are distinguished by obvious characteristics, such as physical appearance (e.g., colour, texture, landscape position), performance (e.g., production capability, flooding), and accompanying vegetation. A vernacular distinction familiar to many is classifying texture as heavy or light.

Light soil content and better structure, take less effort to turn and cultivate. Contrary to popular belief, light soils do not weigh less than heavy soils on an air dry basis nor do they have more porosity.

History

Vasily Dokuchaev, a Russian geologist, geographer and early soil scientist, is credited with identifying soil as a resource whose distinctness and complexity deserved to be separated conceptually from geology and crop production and treated as a whole. Previously, soil had been considered a product of chemical transformations of rocks, a dead substrate from which plants derive nutritious elements.

Soil and bedrock were in fact equated. Dokuchaev considers the soil as a natural body having its own genesis and its own history of development, a body with complex and multiform processes taking place within it. The soil is considered as different from bedrock. The latter becomes soil under the influence of a series of soil-formation factors (climate, vegetation, country, relief and age). According to him, soil should be called the "daily" or outward horizons of rocks regardless of the type; they are changed naturally by the common effect of water, air and various kinds of living and dead organisms.

A 1914 encyclopedic definition: "the different forms of earth on the surface of the rocks, formed by the breaking down or weathering of rocks". serves to illustrate the historic view of soil which persisted from the 19th century. Dokuchaev's late 19th century soil concept developed in the 20th century to one of soil as earthy material that has been altered by living processes. A corollary concept is that soil without a living component is simply a part of earth's outer layer. Further refinement of the soil concept is occurring in view of an appreciation of energy transport and transformation within soil. The term is popularly applied to the material on the surface of the Earth's moon and Mars, a usage acceptable within a portion of the scientific community. Accurate to this modern understanding of soil is Nikiforoff's 1959 definition of soil as the "excited skin of the sub aerial part of the earth's crust".

Areas of Practice

Academically, soil scientists tend to be drawn to one of five areas of specialization: microbiology, pedology, edaphology, physics or chemistry. Yet the work specifics are very much dictated by the challenges facing our civilization's desire to sustain the land that supports it, and the distinctions between the sub-disciplines of soil science often blur in the process. Soil science professionals commonly stay current in soil chemistry, soil physics, soil microbiology, pedology, and applied soil science in related disciplines

One interesting effort drawing in soil scientists in the USA as of 2004 is the Soil Quality Initiative. Central to the Soil Quality Initiative is developing indices of soil health and then monitoring them in a way that gives us long term (decade-to-decade) feedback on our performance as stewards of the planet. The effort includes understanding the functions of soil microbiotic crusts and exploring the potential to sequester atmospheric carbon in soil organic matter. The concept of soil quality, however, has not been without its share of controversy and criticism, including critiques by Nobel Laureate Norman Borlaug and World Food Prize Winner Pedro Sanchez.

A more traditional role for soil scientists has been to map soils. Most every area in the United States now has a published soil survey, which includes interpretive tables as to how soil properties support or limit activities and uses. An internationally accepted soil taxonomy allows uniform communication of soil characteristics and functions. National and international soil survey efforts have given the profession unique insights into landscape scale functions. The landscape functions that soil scientists are called upon to address in the field seem to fall roughly into six areas:

- Land-based treatment of wastes
  - Septic system
  - Manure
  - Municipal biosolids
  - Food and fibre processing waste
- Identification and protection of environmentally critical areas
  - Sensitive and unstable soils
  - Wetlands
  - Unique soil situations that support valuable habitat, and ecosystem diversity
- Management for optimum land productivity
  - Silviculture
  - Agronomy
    - Nutrient management
    - Water management

    - o Native vegetation
    - o Grazing
- Management for optimum water quality
    - o Stormwater management
    - o Sediment and erosion control
- Remediation and restoration of damaged lands
    - o Mine reclamation
    - o Flood and storm damage
    - o Contamination
- Sustainability of desired uses
    - o Soil conservation

There are also practical applications of soil science that might not be apparent from looking at a published soil survey.

- Radiometric dating: specifically a knowledge of local pedology is used to date prior activity at the site
    - o Stratification (archeology) where soil formation processes and preservative qualities can inform the study of archaeological sites
    - o Geological phenomena
        - § Landslides
        - § Active faults
- Altering soils to achieve new uses
    - o Vitrification to contain radioactive wastes
    - o Enhancing soil microbial capabilities in degrading contaminants (bioremediation).
    - o Carbon sequestration
    - o Environmental soil science
- Pedology
    - o Soil genesis
    - o Pedometrics
    - o Soil morphology
        - § Soil micromorphology
    - o Soil classification
        - § USDA soil taxonomy
- Soil biology

  - Soil microbiology
- Soil chemistry
  - Soil biochemistry
  - Soil mineralogy
- Soil physics
  - Pedotransfer function
  - Soil mechanics and engineering
- Soil hydrology, hydropedology

Fields of Application in Soil Science

- Soil survey
- Soil management
- Standard methods of analysis
- Soil fertility / Nutrient management
- Ecosystem studies
- Climate change
- Watershed and wetland studies
- Pedotransfer function

Related Disciplines

- Agricultural sciences
  - Agrophysics science
  - Irrigation management
- Anthropology
  - archaeological stratigraphy
- Environmental science
  - Landscape ecology
- Physical geography
  - Geomorphology
- Geology
  - Biogeochemistry
  - Geomicrobiology
- Hydrology
  - Hydrogeology
- Waste management
- Wetland science

Edaphology

Edaphology is one of two main divisions of soil science, the other being pedology. Edaphology is concerned with the influence of soils on living things, particularly plants. The term is also applied to the study of how soil influences man's use of land for plant growth as well as man's overall use of the land. General subfields within edaphology are agricultural soil science (known by the term agrology in some regions) and environmental soil science. (Pedology deals with pedogenesis, soil morphology, and soil classification.) In Russia, edaphology is considered equivalent to pedology, but is recognized to have an applied sense consistent with agrophysics and agrochemistry outside of Russia.

History

Xenophon (431–355 BC), and Cato (234–149 BC), were early edaphologists. Xenophon noted the beneficial effect of turning a cover crop into the earth. Cato wrote De Agri Cultura ("On Farming") which recommended tillage, crop rotation and the use of legumes in the rotation to build soil nitrogen. He also devised the first soil capability classification for specific crops. Jan Baptist van Helmont (1577–1644) performed a famous experiment, growing a willow tree in a pot of soil and supplying only rainwater for five years. The weight gained by the tree was greater than the weight loss of the soil. He concluded that the willow was made of water. Although only partly correct, his experiment reignited interest in edaphology.

Areas of Study

Agricultural Soil Science

Agricultural soil science is the application of soil chemistry, physics, and biology dealing with the production of crops. In terms of soil chemistry, it places particular emphasis on plant nutrients of importance to farming and horticulture, especially with regard to soil fertility and fertilizer components. Physical edaphology is strongly associated with crop irrigation and drainage. Soil husbandry is a strong tradition within agricultural soil science. Beyond preventing soil erosion and degradation in cropland, soil husbandry seeks to sustain the agricultural soil resource though the use of soil conditioners and cover crops.

Environmental Soil Science

Environmental soil science studies our interaction with the pedosphere on beyond crop production. Fundamental and applied aspects the field address vadose zone functions, septic drain field site assessment and function, land treatment of wastewater, stormwater, erosion control, soil contamination with metals and pesticides, remediation of contaminated soils, restoration of wetlands, soil degradation, and environmental nutrient management. It also studies soil in the context of land use planning, global warming, and acid rain.

Tropical Geography

Tropical geography refers to the study of places and people in the tropics. When it first emerged as a discipline, tropical geography was closely associated with imperialism and colonial expansion of the European empires as contributing scholars tended to portray the tropical places as "primitive" and people "uncivilised" and "inferior". A wide range of subjects has been discussed within the sub-field during late 18th to early 20th century including zoology, climatology, geomorphology, economics and cultural studies.

The discipline is now more commonly known as development geography as colonization had been replaced by economic development as the main ideological driver of international and global interactions since the 1950s. Today, many scholars continue to use the term tropical geography to contest the determinism embedded in the term and de-exoticise the tropical countries and their inhabitants.

Origins

The origins of tropical geography can be traced back to as early as the fifteenth century when Columbus first discovered the Caribbean islands in tropical America. Subsequent writings of European explorers, merchants, naturalists, colonists and settlers who travelled to and lived in the tropics were the main sources of the study.

Alexander von Humboldt, Joseph Dalton Hooker, Charles Darwin and Alfred Russel Wallace are some of the significant contributors. It is argued that it is due to their academic reputation and scientific approaches tropical geography was consolidated into an academic discipline widely studied in Europe in spite of the region's vast differences in vegetation, wild lives, climate, geology and culture.

Troubled Representations

The discourse on the tropics and their inhabitants have evolved overtime in response to changing patterns of Europe's engagements in the tropics. At first, explorers depicted the tropics as lands of great abundance and organic fertility with exotic flora and animals which they have never seen in the temperate. However, this view was challenged when tropical dwellers were taken into consideration.

A variety of environmental determinism emerged out of the sub-field as colonists and naturalists started representing temperate and tropical people with binaries like "progressive vs. backward," "civilised vs. primitive," "hard working vs. lazy" and "superior vs. inferior." Race, an invented concept, was convenient and readily applied in attempts to "[link] climatic variation closely to the supposed division of the human species into different 'races'".

As activities of the European empires diversified in 19th century, travellers and settlers who had experienced deadly tropical diseases and conflicts with the local peoples forged another representation of the tropical world as a place full of "dangers" and "horrors" to mankind. The fertile lands of the tropics were then interpreted as to have set obstacles for human morality and physical well-being preventing their inhabitants from technical, philosophical and artistic innovation. This dramatized and pessimistic representation reinforced Europe's superior position and enhanced the depiction of the tropics as an exotic other to the temperate world.

Whether tropical geographers found the tropical places and people abundant and dynamic or deadly and barbaric, they understood them as inferior to the temperate and great Western civilizations. As criticized by Edward Said in his famous work *Orientalism,* the literature of tropical geography, like writings on the Orient, served the interests of European scholars who were living in the temperate world to create an exotic other which in turn helped define themselves.

Today's Tropical Geography

Until mid 20th century, the imperialist, racist and Euro-centric version of tropical geography was still flourishing as influential works were still being published like *Les Pays Tropicaux* by geomorphologist Pierre Gourou. From 1950's onward, development geography replaced tropical geography as the sub-field of geography. Consequently, the studied regions were given new terminologies such as the "Third World" and the "global South."

Critical geographers argue the replacement of tropical geography by development geography marks the historical turning point of international intervention strategy from colonisation to economic development. Though morphed into a different discipline, the ideological roots of tropical geography—superiority, progress, civilization and technological advancement of the West which originated and matured in the temperate zone—carry on and become the building blocks of mainstream economic development theories.

Paralleled with the rise of critical approaches to other sub-fields of geography, development geography and tropical geography as academic disciplines have seen a movement away from the mainstream economistic and deterministic view of the tropical world in the 1970s. Geographers now attend to the influences imperialism, racism and Euro-centrism have had on tropical geography while attempting to bring class, gender, race and religion into the broader picture to better understand the tropical world and its inhabitants.

Today the academic journal *Singapore Journal of Tropical Geography* continues to be a forum for tropical geographers to introduce and present new research and critique existing literature on the tropical world and people.

Different from the historic absence of voices from the tropics, now many of the contributors of the *Singapore Journal of Tropical Geography* are of tropical origins and study tropical countries and their citizens with more holistic and inclusive approaches.

Phytogeography

Phytogeography (from Greek phyto = plant and geography meaning also distribution) is the branch of biogeography that is concerned with the geographic distribution of plant species and their influence on the earth's surface.

Phytogeography is concerned with all aspects of plant distribution, from the controls on the distribution of individual species ranges (at both large and small scales, see species distribution) to the factors that govern the composition of entire communities and floras. Geobotany, by contrast, focuses on the geographic space's influence on plants.

The basic data elements of phytogeography are occurrence records (presence or absence of a species) with operational geographic units such as political units or geographical coordinates. These data are often used to construct phytogeographic provinces (floristic provinces) and elements.

The questions and approaches in phytogeography are largely shared with zoogeography, except zoogeography is concerned with animal distribution rather than plant distribution. The term phytogeography itself suggests a broad meaning. How the term is actually applied by practicing scientists is apparent in the way periodicals use the term. The *American Journal of Botany*, a monthly primary research journal, frequently publishes a section titled "Systematics, Phytogeography, and Evolution." Topics covered in the *American Journal of Botany*'s "Systematics and Phytogeography" section include phylogeography, distribution of genetic variation and, historical biogeography, and general plant species distribution patterns. Biodiversity patterns are not heavily covered.

History

Phytogeography has a long history. One of the subjects earliest proponents was Prussian naturalist Alexander Von Humboldt, who is often referred to as the "father of phytogeography". Von Humboldt advocated a quantitative approach to phytogeography that has characterized modern plant geography. Gross patterns of the distribution of plants became apparent early on in the study of plant geography. For example, Alfred Russel Wallace, co-discoverer of the principle of natural selection, discussed the Latitudinal gradients in species diversity, a pattern observed in other organisms as well. Much research effort in plant geography has since then has been devoted to understanding this pattern and describing it in more detail.

Research in plant geography has also been directed to understanding the patterns of adaptation of species to the environment. This is done chiefly by describing geographical patterns of trait/environment relationships. These patterns termed ecogeographical rules when applied to plants represent another area of phytogeography. Recently, a new field termed macroecology has developed, which focuses on broad-scale (in both time and space) patterns and phenomena in ecology. Macroecology focuses as much on other organisms as plants.

Branches of Biology Relevant to Phytogeography

Phytogeography is part of a more general science known as biogeography. Phytogeographers are concerned with patterns and process in plant distribution. Most of the major questions and kinds of approaches taken to answer such questions are held in common between phyto- and zoogeographers. Phytogeography is often divided into two main branches: Ecological phytogeography and Historical phytogeography. The former investigates the role of current day biotic and abiotic interactions in influencing plant distributions; the latter are concerned with historical reconstruction of the origin, dispersal, and extinction of taxa. Floristics is a study of the flora of some territory or area. Traditional phytogeography concerns itself largely with floristics and floristic classification.

Landform

A landform in the earth sciences and geology sub-fields, comprises a geomorphological unit, and is largely defined by its surface form and location in the landscape, as part of the terrain, and as such, is typically an element of topography. Landform elements also include seascape and oceanic waterbody interface features such as bays, peninsulas, seas and so forth, including sub-aqueous terrain features such as mid-ocean ridges, volcanoes, and the great ocean basins.

Physical Characteristics

Landforms are categorised by characteristic physical attributes such as elevation, slope, orientation, stratification, rock exposure, and soil type. Gross physical features or landforms include intuitive elements such as berms, mounds, hills, ridges, cliffs, valleys, rivers, peninsulas and numerous other structural and size-scaled (i.e. ponds vs. lakes, hills vs. mountains) elements including various kinds of inland and oceanic waterbodies and sub-surface features.

Hierarchy of Classes

Oceans and continents exemplify the highest-order landforms. Landform elements are parts of a high-order landforms that can be further identified and systematically given a cohesive definition such as hill-tops, shoulders, saddles, foreslopes and backslopes.

Some generic landform elements including: pits, peaks, channels, ridges, passes, pools and plains, may be extracted from a digital elevation model using some automated techniques where the data has been gathered by modern satellites and stereoscopic aerial surveillance cameras. Until recently, compiling the data found in such data sets required time consuming and expensive techniques of many man-hours.

Terrain (or *relief*) is the third or vertical dimension of *land surface*. Topography is the study of terrain, although the word is often used as a synonym for relief itself. When relief is described underwater, the term bathymetry is used. In cartography, many different techniques are used to describe relief, including contour lines and TIN (Triangulated irregular network). Elementary landforms (segments, facets, relief units) are the smallest homogeneous divisions of the land surface, at the given scale/resolution.

These are areas with relatively homogenous morphometric properties, bounded by lines of discontinuity. A plateau or a hill can be observed at various scales ranging from few hundred metres to hundreds of kilometres. Hence, the spatial distribution of landforms is often scale-dependent as is the case for soils and geological strata.

A number of factors, ranging from plate tectonics to erosion and deposition, can generate and affect landforms. Biological factors can also influence landforms— for example, note the role of vegetation in the development of dune systems and salt marshes, and the work of corals and algae in the formation of coral reefs.

Landforms do not include man-made features, such as canals, ports and many harbours; and geographic features, such as deserts, forests, grasslands, and impact craters. Many of the terms are not restricted to refer to features of the planet Earth, and can be used to describe surface features of other planets and similar objects in the Universe. Examples are mountains, hills, polar caps, and valleys, which are found on all of the terrestrial planets.

Universe

The universe is commonly defined as the totality of existence, including planets, stars, galaxies, the contents of intergalactic space, and all matter and energy. Definitions and usage vary and similar terms include the *cosmos*, the *world* and *nature*. Scientific observation of earlier stages in the development of the universe, which can be seen at great distances, suggests that the universe has been governed by the same physical laws and constants throughout most of its extent and history.

There are various multiverse hypotheses, in which physicists have suggested that the universe might be one among many universes that likewise exist. The farthest distance that it is theoretically possible for humans to see is described as the observable universe. Observations have shown that the universe appears to be expanding at an accelerating rate, and a number of models have arisen to predict its ultimate fate.

History

Observational History

Throughout recorded history, several cosmologies and cosmogonies have been proposed to account for observations of the universe. The earliest quantitative geocentric models were developed by the ancient Greek philosophers. Over the centuries, more precise observations and improved theories of gravity led to Copernicus's heliocentric model and the Newtonian model of the Solar System, respectively.

Further improvements in astronomy led to the realization that the Solar System is embedded in a galaxy composed of billions of stars, the Milky Way, and that other galaxies exist outside it, as far as astronomical instruments can reach. Careful studies of the distribution of these galaxies and their spectral lines have led to much of modern cosmology. Discovery of the red shift and cosmic microwave background radiation revealed that the universe is expanding and apparently had a beginning.

History of the Universe

According to the prevailing scientific model of the universe, known as the Big Bang, the universe expanded from an extremely hot, dense phase called the Planck epoch, in which all the matter and energy of the observable universe was concentrated. Since the Planck epoch, the universe has been expanding to its present form, possibly with a brief period (less than 10 seconds) of cosmic inflation. Several independent experimental measurements support this theoretical expansion and, more generally, the Big Bang theory.

Recent observations indicate that this expansion is accelerating because of dark energy, and that most of the matter in the universe may be in a form which cannot be detected by present instruments, called dark matter. The common use of the "dark matter" and "dark energy" placeholder names for the unknown entities purported to account for about 95% of the mass-energy density of the universe demonstrates the present observational and conceptual shortcomings and uncertainties concerning the nature and ultimate fate of the universe.

Current interpretations of astronomical observations indicate that the age of the universe is 13.75 ± 0.17 billion years, (whereas the decoupling of light and matter, see CMBR, happened already 380,000 years after the Big Bang), and that the diametre of the observable universe is at least 93 billion light years or 8.80×10^{26} metres. According to general relativity, space can expand faster than the speed of light, although we can view only a small portion of the universe due to the limitation imposed by light speed. Since we cannot observe space beyond the limitations of light (or any electromagnetic radiation), it is uncertain whether the size of the universe is finite or infinite.

Etymology, Synonyms and Definitions

The word *universe* derives from the Old French word *Univers*, which in turn derives from the Latin word *universum*. The Latin word was used by Cicero and later Latin authors in many of the same senses as the modern English word is used. The Latin word derives from the poetic contraction *Unvorsum* — first used by Lucretius in Book IV (line 262) of his *De rerum natura* (*On the Nature of Things*) — which connects *un, uni* (the combining form of *unus*, or "one") with *vorsum, versum* (a noun made from the perfect passive participle of *vertere*, meaning "something rotated, rolled, changed").

An alternative interpretation of *unvorsum* is "everything rotated as one" or "everything rotated by one". In this sense, it may be considered a translation of an earlier Greek word for the universe, originally used to describe a course of a meal, the food being carried around the circle of dinner guests. This Greek word refers to celestial spheres, an early Greek model of the universe. Regarding Plato's Metaphor of the sun, Aristotle suggests that the rotation of the sphere of fixed stars inspired by the prime mover, motivates, in turn, terrestrial change via the Sun. Careful astronomical and physical measurements (such as the Foucault pendulum) are required to prove the Earth rotates on its axis. The broadest definition of the universe is found in *De divisione naturae* by the medieval philosopher and theologian Johannes Scotus Eriugena, who defined it as simply everything: everything that is created and everything that is not created.

Definition as Reality

More customarily, the universe is defined as everything that exists, (has existed, and will exist). According to our current understanding, the universe consists of three principles: spacetime, forms of energy, including momentum and matter, and the physical laws that relate them.

Definition as Connected Space-time

It is possible to conceive of disconnected space-times, each existing but unable to interact with one another. An easily visualized metaphor is a group of separate soap bubbles, in which observers living on one soap bubble cannot interact with those on other soap bubbles, even in principle. According to one common terminology, each "soap bubble" of space-time is denoted as a universe, whereas our particular space-time is denoted as *the universe*, just as we call our moon *the Moon*.

The entire collection of these separate space-times is denoted as the multiverse. In principle, the other unconnected universes may have different dimensionalities and topologies of space-time, different forms of matter and energy, and different physical laws and physical constants, although such possibilities are currently speculative.

Definition as Observable Reality

According to a still-more-restrictive definition, the universe is everything within our connected space-time that could have a chance to interact with us and vice versa. According to the general theory of relativity, some regions of space may never interact with ours even in the lifetime of the universe, due to the finite speed of light and the ongoing expansion of space. For example, radio messages sent from Earth may never reach some regions of space, even if the universe would live forever; space may expand faster than light can traverse it. It is worth emphasizing that those distant regions of space are taken to exist and be part of reality as much as we are; yet we can never interact with them.

The spatial region within which we can affect and be affected is denoted as the observable universe. Strictly speaking, the observable universe depends on the location of the observer. By travelling, an observer can come into contact with a greater region of space-time than an observer who remains still, so that the observable universe for the former is larger than for the latter. Nevertheless, even the most rapid traveller will not be able to interact with all of space. Typically, the observable universe is taken to mean the universe observable from our vantage point in the Milky Way Galaxy.

Size, Age, Contents, Structure and Laws

The universe is immensely large and possibly infinite in volume. The region visible from Earth (the observable universe) is a sphere with a radius of about 46 billion light years, based on where the expansion of space has taken the most distant objects observed. For comparison, the diametre of a typical galaxy is only 30,000 light-years, and the typical distance between two neighbouring galaxies is only 3 million light-years. As an example, our Milky Way Galaxy is roughly 100,000 light years in diametre, and our nearest sister galaxy, the Andromeda Galaxy, is located roughly 2.5 million light years away. There are probably more than 100 billion (10^{11}) galaxies in the observable universe. Typical galaxies range from dwarfs with as few as ten million (10^7) stars up to giants with one trillion (10^{12}) stars, all orbiting the galaxy's center of mass.

A 2010 study by astronomers estimated that the observable universe contains 300 sextillion (3×10^{23}) stars.The observable matter is spread homogeneously (*uniformly*) throughout the universe, when averaged over distances longer than 300 million light-years. However, on smaller length-scales, matter is observed to form "clumps", i.e., to cluster hierarchically; many atoms are condensed into stars, most stars into galaxies, most galaxies into clusters, superclusters and, finally, the largest-scale structures such as the Great Wall of galaxies.

The observable matter of the universe is also spread *isotropically*, meaning that no direction of observation seems different from any other; each region of the sky has roughly the same content. The universe is also bathed in a highly isotropic microwave radiation that corresponds to a thermal equilibrium blackbody spectrum of roughly 2.725 kelvin. The hypothesis that the large-scale universe is homogeneous and isotropic is known as the cosmological principle, which is supported by astronomical observations.

The present overall density of the universe is very low, roughly $9.9 \times 10^{"30}$ grams per cubic centimetre. This mass-energy appears to consist of 73% dark energy, 23% cold dark matter and 4% ordinary matter. Thus the density of atoms is on the order of a single hydrogen atom for every four cubic metres of volume. The properties of dark energy and dark matter are largely unknown. Dark matter gravitates as ordinary matter, and thus works to slow the expansion of the universe; by contrast, dark energy accelerates its expansion.

The most precise estimate of the universe's age is 13.72±0.12 billion years old, based on observations of the cosmic microwave background radiation. Independent estimates (based on measurements such as radioactive dating) agree at 13–15 billion years. The universe has not been the same at all times in its history; for example, the relative populations of quasars and galaxies have changed and space itself appears to have expanded.

This expansion accounts for how Earth-bound scientists can observe the light from a galaxy 30 billion light years away, even if that light has travelled for only 13 billion years; the very space between them has expanded. This expansion is consistent with the observation that the light from distant galaxies has been redshifted; the photons emitted have been stretched to longer wavelengths and lower frequency during their journey. The rate of this spatial expansion is accelerating, based on studies of Type Ia supernovae and corroborated by other data.

The relative fractions of different chemical elements — particularly the lightest atoms such as hydrogen, deuterium and helium — seem to be identical throughout the universe and throughout its observable history. The universe seems to have much more matter than antimatter, an asymmetry possibly related to the observations of CP violation.

The universe appears to have no net electric charge, and therefore gravity appears to be the dominant interaction on cosmological length scales. The universe also appears to have neither net momentum nor angular momentum. The absence of net charge and momentum would follow from accepted physical laws (Gauss's law and the non-divergence of the stress-energy-momentum pseudotensor, respectively), if the universe were finite.

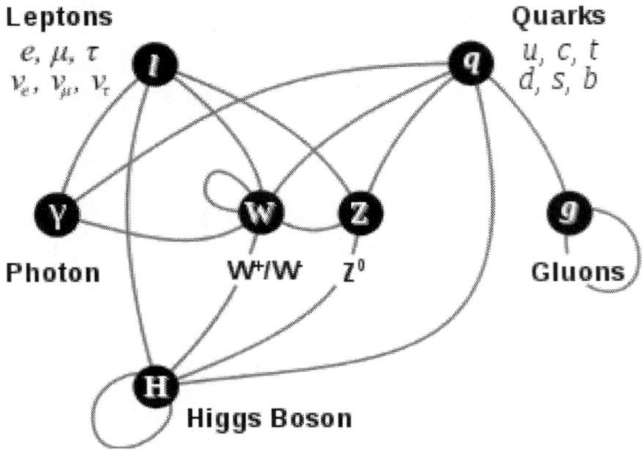

The elementary particles from which the universe is constructed. Six leptons and six quarks comprise most of the matter; for example, the protons and neutrons of atomic nuclei are composed of quarks, and the ubiquitous electron is a lepton. These particles interact via the gauge bosons shown in the middle row, each corresponding to a particular type of gauge symmetry. The Higgs boson is believed to confer mass on the particles with which it is connected. The graviton, a supposed gauge boson for gravity, is not shown.

The universe appears to have a smooth space-time continuum consisting of three spatial dimensions and one temporal (time) dimension. On the average, space is observed to be very nearly flat (close to zero curvature), meaning that Euclidean geometry is experimentally true with high accuracy throughout most of the Universe. Spacetime also appears to have a simply connected topology, at least on the length-scale of the observable universe. However, present observations cannot exclude the possibilities that the universe has more dimensions and that its spacetime may have a multiply connected global topology, in analogy with the cylindrical or toroidal topologies of two-dimensional spaces.

The universe appears to behave in a manner that regularly follows a set of physical laws and physical constants. According to the prevailing

Standard Model of physics, all matter is composed of three generations of leptons and quarks, both of which are fermions. These elementary particles interact via at most three fundamental interactions: the electroweak interaction which includes electromagnetism and the weak nuclear force; the strong nuclear force described by quantum chromodynamics; and gravity, which is best described at present by general relativity.

The first two interactions can be described by renormalized quantum field theory, and are mediated by gauge bosons that correspond to a particular type of gauge symmetry. A renormalized quantum field theory of general relativity has not yet been achieved, although various forms of string theory seem promising.

The theory of special relativity is believed to hold throughout the universe, provided that the spatial and temporal length scales are sufficiently short; otherwise, the more general theory of general relativity must be applied. There is no explanation for the particular values that physical constants appear to have throughout our universe, such as Planck's constant h or the gravitational constant G. Several conservation laws have been identified, such as the conservation of charge, momentum, angular momentum and energy; in many cases, these conservation laws can be related to symmetries or mathematical identities.

Fine Tuning

It appears that many of the properties of the universe have special values in the sense that a universe where these properties only differ slightly would not be able to support intelligent life. Not all scientists agree that this fine-tuning exists. In particular, it is not known under what conditions intelligent life could form and what form or shape that would take. A relevant observation in this discussion is that for an observer to exist to observe fine-tuning, the universe must be able to support intelligent life. As such the conditional probability of observing a universe that is fine-tuned to support intelligent life is 1. This observation is known as the anthropic principle and is particularly relevant if the creation of the universe was probabilistic or if multiple universes with a variety of properties exist.

Historical Models

Many models of the cosmos (cosmologies) and its origin (cosmogonies) have been proposed, based on the then-available data and conceptions of the universe. Historically, cosmologies and cosmogonies were based on narratives of gods acting in various ways. Theories of an impersonal universe governed by physical laws were first proposed by the Greeks and Indians.

Over the centuries, improvements in astronomical observations and theories of motion and gravitation led to ever more accurate descriptions of the universe. The modern era of cosmology began with Albert Einstein's 1915 general theory of relativity, which made it possible to quantitatively predict the origin, evolution, and conclusion of the universe as a whole. Most modern, accepted theories of cosmology are based on general relativity and, more specifically, the predicted Big Bang; however, still more careful measurements are required to determine which theory is correct.

Creation

Many cultures have stories describing the origin of the world, which may be roughly grouped into common types. In one type of story, the world is born from a world egg; such stories include the Finnish epic poem *Kalevala*, the Chinese story of Pangu or the Indian Brahmanda Purana. In related stories, the creation idea is caused by a single entity emanating or producing something by him- or herself, as in the Tibetan Buddhism concept of Adi-Buddha, the ancient Greek story of Gaia (Mother Earth), the Aztec goddess Coatlicue myth, the ancient Egyptian god Atum story, or the Genesis creation narrative. In another type of story, the world is created from the union of male and female deities, as in the Maori story of Rangi and Papa.

In other stories, the universe is created by crafting it from pre-existing materials, such as the corpse of a dead god — as from Tiamat in the Babylonian epic Enuma Elish or from the giant Ymir in Norse mythology – or from chaotic materials, as in Izanagi and Izanami in Japanese mythology. In other stories, the universe emanates from fundamental principles, such as Brahman and Prakrti, the creation myth of the Serers, or the yin and yang of the Tao.

Philosophical Models

From the 6th century BCE, the pre-Socratic Greek philosophers developed the earliest known philosophical models of the universe. The earliest Greek philosophers noted that appearances can be deceiving, and sought to understand the underlying reality behind the appearances. In particular, they noted the ability of matter to change forms (e.g., ice to water to steam) and several philosophers proposed that all the apparently different materials of the world are different forms of a single primordial material, or arche.

The first to do so was Thales, who proposed this material is Water. Thales' student, Anaximander, proposed that everything came from the limitless apeiron. Anaximenes proposed Air on account of its perceived attractive and repulsive qualities that cause the arche to condense or dissociate into different forms. Anaxagoras, proposed the principle of Nous (Mind). Heraclitus proposed fire (and spoke of logos). Empedocles proposed the elements: earth, water, air and fire.

His four element theory became very popular. Like Pythagoras, Plato believed that all things were composed of number, with the Empedocles' elements taking the form of the Platonic solids. Democritus, and later philosophers—most notably Leucippus—proposed that the universe was composed of indivisible atoms moving through void (vacuum). Aristotle did not believe that was feasible because air, like water, offers resistance to motion. Air will immediately rush in to fill a void, and moreover, without resistance, it would do so indefinitely fast.

Although Heraclitus argued for eternal change, his quasi-contemporary Parmenides made the radical suggestion that all change is an illusion, that the true underlying reality is eternally unchanging and of a single nature. Parmenides' theory seemed implausible to many Greeks, but his student Zeno of Elea challenged them with several famous paradoxes. Aristotle responded to these paradoxes by developing the notion of a potential countable infinity, as well as the infinitely divisible continuum. Unlike the eternal and unchanging cycles of time, he believed the world was bounded by the celestial spheres, and thus magnitude was only finitely multiplicative. The Indian philosopher Kanada, founder of the Vaisheshika school, developed a theory of atomism and proposed that light and heat were varieties of the same substance. In the 5th century AD, the Buddhist atomist philosopher Dignâga proposed atoms to be point-sized, durationless, and made of energy. They denied the existence of substantial matter and proposed that movement consisted of momentary flashes of a stream of energy.

The theory of temporal finitism was inspired by the doctrine of Creation shared by the three Abrahamic religions: Judaism, Christianity and Islam. The Christian philosopher, John Philoponus, presented the philosophical arguments against the ancient Greek notion of an infinite past and future.

Philoponus' arguments against an infinite past were used by the early Muslim philosopher, Al-Kindi (Alkindus); the Jewish philosopher, Saadia Gaon (Saadia ben Joseph); and the Muslim theologian, Al-Ghazali (Algazel). Borrowing from Aristotle's *Physics* and *Metaphysics*, they employed two logical arguments against an infinite past, the first being the "argument from the impossibility of the existence of an actual infinite", which states:

"An actual infinite cannot exist."

"An infinite temporal regress of events is an actual infinite."

"∴ An infinite temporal regress of events cannot exist."

The second argument, the "argument from the impossibility of completing an actual infinite by successive addition", states:

"An actual infinite cannot be completed by successive addition."

"The temporal series of past events has been completed by successive addition."

" The temporal series of past events cannot be an actual infinite."

Both arguments were adopted by Christian philosophers and theologians, and the second argument in particular became more famous after it was adopted by Immanuel Kant in his thesis of the first antinomy concerning time.

Astronomical Models

Astronomical models of the universe were proposed soon after astronomy began with the Babylonian astronomers, who viewed the universe as a flat disk floating in the ocean, and this forms the premise for early Greek maps like those of Anaximander and Hecataeus of Miletus.

Later Greek philosophers, observing the motions of the heavenly bodies, were concerned with developing models of the universe based more profoundly on empirical evidence. The first coherent model was proposed by Eudoxus of Cnidos. According to Aristotle's physical interpretation of the model, celestial spheres eternally rotate with uniform motion around a stationary Earth.

Normal matter, is entirely contained within the terrestrial sphere. This model was also refined by Callippus and after concentric spheres were abandoned, it was brought into nearly perfect agreement with astronomical observations by Ptolemy. The success of such a model is largely due to the mathematical fact that any function (such as the position of a planet) can be decomposed into a set of circular functions (the Fourier modes).

Other Greek scientists, such as the Pythagorean philosopher Philolaus postulated that at the center of the universe was a "central fire" around which the Earth, Sun, Moon and Planets revolved in uniform circular motion. The Greek astronomer Aristarchus of Samos was the first known individual to propose a heliocentric model of the universe. Though the original text has been lost, a reference in Archimedes' book The Sand Reckoner describes Aristarchus' heliocentric theory. Archimedes wrote: (translated into English)

You King Gelon are aware the 'universe' is the name given by most astronomers to the sphere the center of which is the center of the Earth, while its radius is equal to the straight line between the center of the Sun and the center of the Earth. This is the common account as you have heard from astronomers. But Aristarchus has brought out a book consisting of certain hypotheses, wherein it appears, as a consequence of the assumptions made, that the universe is many times greater than the 'universe' just mentioned.

His hypotheses are that the fixed stars and the Sun remain unmoved, that the Earth revolves about the Sun on the circumference of a circle, the Sun lying in the middle of the orbit, and that the sphere of fixed stars, situated about the same center as the Sun, is so great that the circle in which he supposes the Earth to revolve bears such a proportion to the distance of the fixed stars as the center of the sphere bears to its surface.

Aristarchus thus believed the stars to be very far away, and saw this as the reason why there was no visible parallax, that is, an observed movement of the stars relative to each other as the Earth moved around the Sun. The stars are in fact much farther away than the distance that was generally assumed in ancient times, which is why stellar parallax is only detectable with telescopes. The geocentric model, consistent with planetary parallax, was assumed to be an explanation for the unobservability of the parallel phenomenon, stellar parallax. The rejection of the heliocentric view was apparently quite strong, as the following passage from Plutarch suggests (On the Apparent Face in the Orb of the Moon):

Cleanthes [a contemporary of Aristarchus and head of the Stoics] thought it was the duty of the Greeks to indict Aristarchus of Samos on the charge of impiety for putting in motion the Hearth of the universe [i.e. the earth], . . . supposing the heaven to remain at rest and the earth to revolve in an oblique circle, while it rotates, at the same time, about its own axis.

The only other astronomer from antiquity known by name who supported Aristarchus' heliocentric model was Seleucus of Seleucia, a Hellenistic astronomer who lived a century after Aristarchus. According to Plutarch, Seleucus was the first to prove the heliocentric system through reasoning, but it is not known what arguments he used. Seleucus' arguments for a heliocentric theory were probably related to the phenomenon of tides. According to Strabo (1.1.9), Seleucus was the first to state that the tides are due to the attraction of the Moon, and that the height of the tides depends on the Moon's position relative to the Sun. Alternatively, he may have proved the heliocentric theory by determining the constants of a geometric model for the heliocentric theory and by developing methods to compute planetary positions using this model, like what Nicolaus Copernicus later did in the 16th century. During the Middle Ages, heliocentric models may have also been proposed by the Indian astronomer, Aryabhata, and by the Persian astronomers, Albumasar and Al-Sijzi.

The Aristotelian model was accepted in the Western world for roughly two millennia, until Copernicus revived Aristarchus' theory that the astronomical data could be explained more plausibly if the earth rotated on its axis and if the sun were placed at the center of the universe.

> *" In the center rests the sun. For who would place this lamp of a very beautiful temple in another or better place than this wherefrom it can illuminate everything at the same time? "*
>
> *—Nicolaus Copernicus, in Chapter 10, Book 1 of De Revolutionibus Orbium Coelestrum (1543)*

As noted by Copernicus himself, the suggestion that the Earth rotates was very old, dating at least to Philolaus (c. 450 BC), Heraclides Ponticus (c. 350 BC) and Ecphantus the Pythagorean. Roughly a century before Copernicus, Christian scholar Nicholas of Cusa also proposed that the Earth rotates on its axis in his book, *On Learned Ignorance* (1440).

Aryabhata (476–550), Brahmagupta (598–668), Albumasar and Al-Sijzi, also proposed that the Earth rotates on its axis. The first empirical evidence for the Earth's rotation on its axis, using the phenomenon of comets, was given by Tusi (1201–1274) and Ali Qushji (1403–1474).

This cosmology was accepted by Isaac Newton, Christiaan Huygens and later scientists. Edmund Halley (1720) and Jean-Philippe de

Cheseaux (1744) noted independently that the assumption of an infinite space filled uniformly with stars would lead to the prediction that the nighttime sky would be as bright as the sun itself; this became known as Olbers' paradox in the 19th century. Newton believed that an infinite space uniformly filled with matter would cause infinite forces and instabilities causing the matter to be crushed inwards under its own gravity. This instability was clarified in 1902 by the Jeans instability criterion.

One solution to these paradoxes is the Charlier universe, in which the matter is arranged hierarchically (systems of orbiting bodies that are themselves orbiting in a larger system, *ad infinitum*) in a fractal way such that the universe has a negligibly small overall density; such a cosmological model had also been proposed earlier in 1761 by Johann Heinrich Lambert.

A significant astronomical advance of the 18th century was the realization by Thomas Wright, Immanuel Kant and others of nebulae.The modern era of physical cosmology began in 1917, when Albert Einstein first applied his general theory of relativity to model the structure and dynamics of the universe.

Theoretical Models

Of the four fundamental interactions, gravitation is dominant at cosmological length scales; that is, the other three forces play a negligible role in determining structures at the level of planetary systems, galaxies and larger-scale structures. Since all matter and energy gravitate, gravity's effects are cumulative; by contrast, the effects of positive and negative charges tend to cancel one another, making electromagnetism relatively insignificant on cosmological length scales. The remaining two interactions, the weak and strong nuclear forces, decline very rapidly with distance; their effects are confined mainly to sub-atomic length scales.

General Theory of Relativity

Given gravitation's predominance in shaping cosmological structures, accurate predictions of the universe's past and future require an accurate theory of gravitation. The best theory available is Albert Einstein's general theory of relativity, which has passed all experimental tests hitherto. However, since rigorous experiments have not been carried out on cosmological length scales, general relativity could conceivably be inaccurate. Nevertheless, its cosmological predictions appear to be consistent with observations, so there is no compelling reason to adopt another theory. General relativity provides a set of ten nonlinear partial differential equations for the spacetime metric (Einstein's field equations) that must be solved from the distribution of mass-energy and momentum throughout the universe. Since these are unknown in exact detail, cosmological models have been based on the cosmological principle, which states that the universe is homogeneous and isotropic.

In effect, this principle asserts that the gravitational effects of the various galaxies making up the universe are equivalent to those of a fine dust distributed uniformly throughout the universe with the same average density. The assumption of a uniform dust makes it easy to solve Einstein's field equations and predict the past and future of the universe on cosmological time scales.

Einstein's field equations include a cosmological constant ($\ddot{E}$), that corresponds to an energy density of empty space. Depending on its sign, the cosmological constant can either slow (negative $\ddot{E}$) or accelerate (positive $\ddot{E}$) the expansion of the universe. Although many scientists, including Einstein, had speculated that $\ddot{E}$ was zero, recent astronomical observations of type Ia supernovae have detected a large amount of "dark energy" that is accelerating the universe's expansion.

Preliminary studies suggest that this dark energy corresponds to a positive $\ddot{E}$, although alternative theories cannot be ruled out as yet. Russian physicist Zel'dovich suggested that $\ddot{E}$ is a measure of the zero-point energy associated with virtual particles of quantum field theory, a pervasive vacuum energy that exists everywhere, even in empty space. Evidence for such zero-point energy is observed in the Casimir effect.

Bibliography

Allen, Kathleen M. S., Stanton W. Green: *Interpreting Space: GIS and Archaeology*, London and New York: Taylor & Francis, 1990.

Armin K.; Gentilli, J. and Fairbridge, R.W.: *Physiographic Diagram of Australia*, The Geological Press, Columbia University, New York, 1951.

Belden C.: *Landscapes of the Sacred: Geography and Narrative in American Spirituality*, New York: Paulist Press, 1988.

Bolstad, P. : *GIS Fundamentals, A first text on Geographic Information Systems, Second Edition*, White Bear Lake, MN, Eider Press, 2005.

Bolstad, P.: *GIS Fundamentals, A first text on Geographic Information Systems, Second Edition*, Eider Press, White Bear Lake, MN, 2005.

Carrillo, J., Lung, Y., and van Tulder, R. : *Cars, Carriers of Regionalism*, Palgrave Macmillan, New York, 2004.

Cronon, William: *Changes in the Land: Indians, Colonists, and the Ecology of New England*, Hill and Wang, New York, 1983.

Davis, Mike: *Ecology of Fear: Los Angeles and the Imagination of Disaster*. New York: Vintage, 1999.

Dibbern, Jens: *The Sourcing of Application Software Services, Empirical Evidence of Cultural, Industry and Functional Differences*, Physica Verlag Heidelberg, 2004.

Elangovan K. : *GIS, Fundamentals, Applications and Population Geography*, New India Publishing Agency, New Delhi 2006.

Harvey, David: *Explanation in Geography*. London: Edward Arnold, 1969.

Harvey, Francis : *A Primer of GIS, Fundamental Geographic and Cartographic Concepts*, The Guilford Press, 2008.

Holden, Joseph: *Introduction to Physical Geography and the Environment*. Prentice-Hall, London, 2004.

Inkpen, Robert: *Science, Philosophy and Physical Geography*, Routledge, London, 2004.

Jackson, Peter: *Maps of Meaning: An Introduction to Cultural Geography*. London and Boston: Unwin Hyman, 1989.

James, T.: *The Best Poor Man's Country: A Geographical Study of Early Southeastern Pennsylvania*, Johns Hopkins Press, Baltimore, 1972.

James, T.: *The Best Poor Man's Country: A Geographical Study of Early Southeastern Pennsylvania*. Baltimore: Johns Hopkins Press, 1972.

King, Anthony D.: *Buildings and Society: Essays on the Social Development of the Built Enviroment*. London: Routledge & Kegan Paul, 1980.

Lewis, Earl. *In Their Own Interests: Race, Class, and Power in Twentieth-Century Norfolk, Virginia*. Berkeley: University of California Press, 1990.

Michael P.: *The Making of the American Landscape*. Boston: Unwin Hyman, 1990.

Michael, H.: *City Form and Natural Process: Towards a New Urban Vernacular,* Croom Helm, London, 1984.

Neil, H.: *The Land of Contrasts: 1880-1901,* The American Culture 5, G. Braziller, New York, 1970.

Neil, H.: *The Land of Contrasts: 1880-1901,* The American Culture 5. New York: G. Braziller, 1970.

Peter, G.: *The Slow Plague: A Geography of the AIDS Pandemic,* Blackwell Publishers, Oxford and Cambridge, MA, 1993.

Peter, G.: *The Slow Plague: A Geography of the AIDS Pandemic*. Oxford and Cambridge, MA: Blackwell Publishers, 1993.

Ray Allen: *The Far Western Frontier, 1830-1860*. The New American Nation Series. New York: Harper, 1956.

Ritchie, J.: *Landscape and Material Life in Franklin County, Massachusetts, 1770-1860*. Knoxville: University of Tennessee Press, 1991.

Stephen, D.: *Fields of Vision: Landscape Imagery and National Identity in England and the United States*. Cambridge: Polity, 1993.

Stephen, D.: *Fields of Vision: Landscape Imagery and National Identity in England and the United States*. Cambridge: Polity, 1993.

Strahler, Alan; Strahler Arthur: *Introducing Physical Geography*, Wiley, New York, 2006.

Tulder, R. : *Cars, Carriers of Regionalism*, Palgrave Macmillan, New York, 2004.

Victor, A. : *Geography of Art and Culture*, Contributions to Economic Analysis, Elsevier, 2004.

Victor, A. : *Geography of Art and Culture*, Contributions to Economic Analysis, Elsevier, 2004.

Watkins, Eric: *The Middle Eastern Environment*, London, The British Society for Middle Eastern Studie, 1995.

Watkins, Eric: *The Middle Eastern Environment*, London, The British Society for Middle Eastern Studie, 1995.

William G. BOWEN : *Performing Arts, The Economic Dilemma,* Thousand Oaks, SAGE, London, 2002.

Wolf, Aaron T.: *Hydropolitics along the Jordan River, Scarce Water and its Impact on the Arab-Israeli Conflict*, NYC, United Nations University Press, 1995.

Index

❑❑❑